Creating Mobile Apps with Appcelerator Titanium

Develop fully-featured mobile applications using a hands-on approach, and get inspired to develop more

Christian Brousseau

BIRMINGHAM - MUMBAI

Creating Mobile Apps with Appcelerator Titanium

First published: October 2013

Production Reference: 1211013

Published by Packt Publishing Ltd.
Livery Place
35 Livery Street
Birmingham B3 2PB, UK.

ISBN 978-1-84951-926-7

www.packtpub.com

Cover Image by Prashant Timappa Shetty (sparkling.spectrum.123@gmail.com)

Credits

Author

Christian Brousseau

Reviewers

Stephen Feather

Jude Osborn

Acquisition Editor

Usha Iyer

Sam Wood

Lead Technical Editor

Balaji Naidu

Technical Editors

Krishnaveni Haridas

Manal Pednekar

Shali Sasidharan

Copy Editors

Gladson Monteiro

Sayanee Mukherjee

Alfida Paiva

Kirti Pai

Project Coordinator

Anugya Khurana

Proofreader

Simran Bhogal

Amisha Green

Indexer

Rekha Nair

Tejal Soni

Graphics

Abhinash Sahu

Yuvraj Mannari

Production Coordinator

Pooja Chiplunkar

Cover Work

Pooja Chiplunkar

About the Author

Christian Brousseau is a proud Canadian who has been developing software for over twenty years. He made his debut in Windows development and then moved on to Java when the Web came along. About thirteen years ago, he had the opportunity to move to France where he worked on major Java Enterprise projects for one of the largest software companies in the world, in different sectors such as banking, insurance, retail, government, and defense.

Over the last few years, he became very enthusiastic about mobile platforms in general and has developed quite a few applications for iOS and Android devices using the Appcelerator Titanium platform. He is very active within the community; being a Titanium-certified application developer and a member of the Titans Evangelist group, he is one of the top 10 contributors on the Appcelerator's **Questions & Answers** forum.

Still living in Paris, he left the enterprise world and created his own company, **Things Are Moving** (http://www.thingsaremoving.com), that specializes in mobile development.

I would like to thank all my friends and colleagues who have given me their support, and all those who were there when I created my very first mobile application (that did not do much at the time I must admit). I'd also like to thank all of the very knowledgeable people who took the time to review my book and gave me more than just review notes. You guys went the extra mile, and I am very grateful for it.

But most of all, I'd like to thank my wife, who gave me her full support in this journey (even during bad times, that support never flinched). Also, to my children who had to see their father writing every weekend instead of spending quality time with them.

To everyone, thank you!

About the Reviewers

Stephen Feather is an Appcelerator Titanium Titan, holding TCMD and TCAD certifications, and a frequent speaker on mobile strategies for small businesses and nonprofit organizations. In 1994, he started his own consulting firm working directly with communication companies such as Netscape, Microsoft, and Oracle in the early days of the Internet. In 1996, he wrote *Javascript by Example*, one of the first publications on the then new scripting language. Over the next 17 years, his firm would grow to become a widely recognized vendor of multimedia software for language learning and providing support to colleges and universities throughout the Southeastern United States.

In 2009, he cofounded *Feather Direct*, recognizing a need for quality mobile application development, reputation management, and SEO services for smaller organizations. He volunteers his time to assist and train a new generation of mobile app developers through online forums and local user groups.

Jude Osborn is an experienced, well-travelled Software Developer, having led and developed projects across the web, mobile, and desktop spaces and in locations spanning the globe. Jude is passionate about new development technologies, and right now is especially giddy about JavaScript frameworks and mobile.

Jude currently works with Google's Creative Lab (on behalf of the web development agency, Potato) on prototypes and experimental software projects. Every morning he walks across Sydney's Pyrmont bridge soaking up the sunshine and looking forward to the next amazing technological challenge.

When he's not cutting code and trying crazy new techs, Jude loves creating Littlest Pet Shop YouTube videos with his daughter, barely managing to compete in Starcraft's gold league with his son, and enjoying a nice glass of wine with his beautiful, amazing, and supportive wife.

www.PacktPub.com

Support files, eBooks, discount offers and more

You might want to visit www.PacktPub.com for support files and downloads related to your book.

Did you know that Packt offers eBook versions of every book published, with PDF and ePub files available? You can upgrade to the eBook version at www.PacktPub.com and as a print book customer, you are entitled to a discount on the eBook copy. Get in touch with us at service@packtpub.com for more details.

At www.PacktPub.com, you can also read a collection of free technical articles, sign up for a range of free newsletters and receive exclusive discounts and offers on Packt books and eBooks.

http://PacktLib.PacktPub.com

Do you need instant solutions to your IT questions? PacktLib is Packt's online digital book library. Here, you can access, read and search across Packt's entire library of books.

Why Subscribe?

- Fully searchable across every book published by Packt
- Copy and paste, print and bookmark content
- On demand and accessible via web browser

Free Access for Packt account holders

If you have an account with Packt at www.PacktPub.com, you can use this to access PacktLib today and view nine entirely free books. Simply use your login credentials for immediate access.

Table of Contents

Preface

Most people are familiar with Titanium Mobile SDK, which allows developers to develop native mobile applications using JavaScript. But Titanium SDK is part of a much larger ecosystem.

Titanium core

What many consider as the main component of Titanium, comprises three major components discussed in the following sections.

Titanium Mobile SDK

Any developer who wants to develop mobile applications using Titanium must use the SDK. It comprises more than 5000 APIs.

The SDK is developed by a company named Appcelerator, but it is an open source project and is released under the Apache License. This means that you can look at the source code and even modify it if you wish.

Titanium studio

Studio is an **Integrated Development Environment** (IDE) based on Aptana Studio (which itself is based on the Eclipse platform). It offers all of the basic features you would expect from an IDE, such as project management, code editor, and some pretty powerful features such as debugging.

While Appcelerator distributes it for free, it is not an open source product. They do, however, provide an SDK if you ever wish to expand some of its features.

Appcelerator analytics

This is an analytics module for developers who want to track the deployment of their app.

The service is free up to a certain point. This means that if you generate a lot of traffic, you will probably have to move over to a paid service plan.

Appcelerator Cloud Services

Appcelerator Cloud Services (ACS) is a cloud service that offers over 20 prebuilt common services you would expect from a cloud service provider. It becomes interesting when it is in very tight integration with the native SDK.

Simply create a new project in Titanium Studio, and boom, you have a cloud-enabled application! This saves enormous efforts in terms of integration. Like analytics, the service is free up to a certain threshold. If you go over that, you will have to move to a paid plan.

Also, ACS is not limited to Titanium applications. In fact, they offer SDK for other platforms and languages as well.

What if this is not enough for my needs?

The day will come where all the APIs offered by Titanium won't be able to fulfill one specific need. Specific as it may be, it might prove vital for your business. Titanium allows more experienced developers to develop extension modules. The name says it all; extension modules extend the capabilities of Titanium while keeping the compatibility with the existing JavaScript code.

This is interesting because some platforms would require you to rewrite the entire application in their respective language (objective C or Java), just to have access to one specific feature of the platform.

Module developers can also hope to recoup their investment by distributing their modules on the Appcelerator marketplace, where other developers can download existing modules (for free or for a fee).

What are the goals of this book?

Each chapter covers the development of a complete mobile application. At the end of each chapter, you will have a working mobile application whether it's for iOS, Android, Blackberry 10, Tizen, or even Mobile web. There is no better way to learn how to do something than building it yourself. Developers can start from any application that has similarities to the one they wish to produce and extend it from there.

Another goal of this book is to introduce some aspects of Titanium that are probably less known by mobile developers.

What this book covers

Chapter 1, Stopwatch (with Lap Counter), guides the reader in creating a standalone Stopwatch application. This first application is pretty straightforward; it covers interactions between user interface elements and data structures.

Chapter 2, Sili, the assistant that just listens, guides the reader in creating a voice recorder application. While Siri interprets what the user is saying, this application simply listens by storing the recordings on the device for later use. It covers media management as well as file system access on the device.

Chapter 3, The To-do List, guides the reader in creating a To-do items management application. It allows the creation and deletion of items, as well as the possibility to mark them as done. All items are persisted in an SQLite database. It covers the integration with an embedded database.

Chapter 4, Interactive E-book for iPad, guides the reader in creating an interactive electronic book (e-book) application with realistic page flipping (like a real book). This is an iPad-specific application in order to benefit from the large screen resolution. It covers the integration and utilization of a native module as well a rich media presentation.

Chapter 5, You've Got to Know When to Hold 'em, guides the reader in creating a standalone Stock Portfolio application. It allows users to organize stocks (price and quantity). From there, the user selects an amount of money he/she wants to gain through his/her investments. The application will periodically retrieve stock prices from the web and calculate the sum of money earned. It will then indicate how far (or close) he/she is from his/her objective. It covers HTTP API calls, property persistence, and custom UI controls.

Chapter 6, JRPG – Second to Last Fantasy, guides the reader in creating a native mobile game. The game shows how players roam around a map from a top view perspective, much similar to a classic Japanese RPG. It covers graphics manipulation and touch control.

Chapter 7, JRPG – Second to Last Fantasy Online, guides the reader in adding online multiplayer functionality to a game. This chapter re-uses the code from the game created in *Chapter 6, JRPG – Second to Last Fantasy*. It covers network interaction, intervals, and testing.

Chapter 8, Social Networks, guides the reader in creating a social application that allows them to update their status on Facebook and Twitter with a single click. It covers application preferences, social network authentication, and integration.

Chapter 9, Marvels of the World Around Us, guides the reader in creating an application that shows online pictures that were taken near the device's current location. It covers location services, web API calls, and photo gallery integration.

Chapter 10, Worldwide Marco Polo, guides the reader in creating a social application that allows users to check in at a location based on the device's location and then share this check in with the world using Appcelerator Cloud Services. It covers location services, cloud integration, and map view.

Appendix, References, contains detailed information about the JavaScript frameworks, libraries, and native extension modules used throughout the book, as well as download locations, and where to find the source code (when applicable).

What you need for this book

You will need a computer with at least 2 GB of memory and a recent version of Windows, Mac OS X, or Ubuntu in order to use Titanium SDK properly. You will also require the Oracle Java Development Kit as well as an install of Node.js to use the **Titanium Command Line Interface (CLI)** tools.

If you are developing an iOS application, you will also need to install the **Xcode Integrated Development Environment (IDE)**. For Android development, you will need the Android SDK provided by Google for free (Titanium Studio provides wizards that can automate the installation).

To develop iOS (iPhone, iPad) applications, you absolutely need a Mac. This limitation is enforced by Apple requiring the use of its own tools. This is not related to Titanium.

Who this book is for

This book is geared towards developers who already have experience with more modern languages and development environments. It is for those who are also familiar with concepts such as **Object Oriented Programming (OOP)**, reusable components, AJAX closures, and so on. This book is also geared for those who are not yet familiar with mobile development, but have a keen interest in this topic.

The book is geared towards existing Titanium developers, who have some (or more) experience with the SDK and the toolset; it is also for those who wish to know more about Titanium and its broad range of capabilities (many of them are not yet very well known). This book can help developers identify how they can extend Titanium's basic set of features by using the extension modules.

This book begins with the basics in the first chapters and it increases throughout the course of the book. The first few chapters will give very detailed step-by-step procedures.

Conventions

In this book, you will find a number of styles of text that distinguish between different kinds of information. Here are some examples of these styles, and an explanation of their meaning.

Code words in text, database table names, folder names, filenames, file extensions, pathnames, dummy URLs, user input, and Twitter handles are shown as follows: " We create the window using the `Ti.Ui.createWindow` function."

A block of code is set as follows:

```
var win = Ti.UI.createWindow({
  backgroundColor: '#ffffff',
  layout: 'vertical'
});
```

When we wish to draw your attention to a particular part of a code block, the relevant lines or items are set in bold:

```
var buttonStartLap = Ti.UI.createButton({
  title: 'GO!',
  color: '#C0BFBF',
  width: '50%',
  height: Ti.UI.FILL,
  backgroundColor: '#727F7F',
  font: {
```

```
    fontSize: '25sp',
    fontWeight: 'bold'
  }
});
```

Any command-line input or output is written as follows:

```
[INFO] 00:01:17:37
[INFO] 00:02:53:19
[INFO] 00:04:06:93
[INFO] 00:06:11:52
```

New terms and **important words** are shown in bold. Words that you see on the screen, in menus or dialog boxes for example, appear in the text like this: "In the **Classic** section, select the **Default Project** template, and then click on **Next >**."

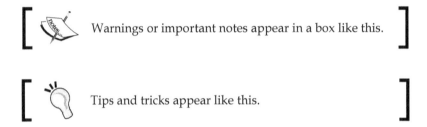

> Warnings or important notes appear in a box like this.

> Tips and tricks appear like this.

Reader feedback

Feedback from our readers is always welcome. Let us know what you think about this book—what you liked or may have disliked. Reader feedback is important for us to develop titles that you really get the most out of.

To send us general feedback, simply send an e-mail to feedback@packtpub.com, and mention the book title via the subject of your message.

If there is a topic that you have expertise in and you are interested in either writing or contributing to a book, see our author guide on www.packtpub.com/authors.

Customer support

Now that you are the proud owner of a Packt book, we have a number of things to help you to get the most from your purchase.

Downloading the example code

You can download the example code files for all Packt books you have purchased from your account at `http://www.packtpub.com`. If you purchased this book elsewhere, you can visit `http://www.packtpub.com/support` and register to have the files e-mailed directly to you.

Downloading the color images of this book

We also provide you a PDF file that has color images of the screenshots/diagrams used in this book. The color images will help you better understand the changes in the output. You can download this file from: `http://www.packtpub.com/sites/default/files/downloads/9267OS_ColoredImages.pdf`.

Errata

Although we have taken every care to ensure the accuracy of our content, mistakes do happen. If you find a mistake in one of our books — maybe a mistake in the text or the code — we would be grateful if you would report this to us. By doing so, you can save other readers from frustration and help us improve subsequent versions of this book. If you find any errata, please report them by visiting `http://www.packtpub.com/submit-errata`, selecting your book, clicking on the **errata submission form** link, and entering the details of your errata. Once your errata are verified, your submission will be accepted and the errata will be uploaded on our website, or added to any list of existing errata, under the Errata section of that title. Any existing errata can be viewed by selecting your title from `http://www.packtpub.com/support`.

Piracy

Piracy of copyright material on the Internet is an ongoing problem across all media. At Packt, we take the protection of our copyright and licenses very seriously. If you come across any illegal copies of our works, in any form, on the Internet, please provide us with the location address or website name immediately so that we can pursue a remedy.

Please contact us at `copyright@packtpub.com` with a link to the suspected pirated material.

We appreciate your help in protecting our authors, and our ability to bring you valuable content.

Questions

You can contact us at `questions@packtpub.com` if you are having a problem with any aspect of the book, and we will do our best to address it.

Stopwatch (with Lap Counter) 1

For our very first application, we will be building a standalone stopwatch. It will provide basic functions such as starting, pausing, and resetting. We will also provide the possibility of saving the lap time and displaying it in a list. This first example is pretty straightforward, but after going through the steps of building the application, we will be able to cover the different interactions between **user interface** (**UI**) elements (for buttons and display items) and data structures (for keeping tabs on our lap times).

By the end of this chapter, you will have learned the following concepts:

- Creating labels, buttons, and views
- Styling them and placing them on the screen
- Creating interactions between UI elements
- Creating a table view (that scrolls) and adding rows to it
- Using Titanium's logging API

Creating our project

Since this is the book's very first mobile application, we will go through the whole project creation process. For those of you who are already familiar with **Titanium Studio**, you can glance over this section in order to get the information you need. As for those of you who haven't delved into Titanium development yet, this will be a good walkthrough. Once you know how to create a project, you will be able to repeat the operation for every application from this book.

Once you have launched Titanium Studio, navigate to **File | New | Mobile Project**. In the **Classic** section, select the **Default Project** template and then click on **Next >**.

Fill out the **Wizard** form with the following information:

Field	Value to enter
Project name	StopWatch
Location	You can follow either one of the ensuing steps: • Create the project in your current workspace directory by selecting the **Use Default Location** checkbox • Create the project in a location of your choice
App Id	com.packtpub.hotshot.stopwatch
Company/Personal URL	http://www.packtpub.com

Field	Value to enter
Titanium SDK Version	By default, the wizard will select the latest version of the Titanium SDK, which is recommended (as of this writing, we are using Version **3.1.3 GA**)
Deployment Targets	Select **iPhone** and **Android**
Cloud Settings	Uncheck the **Cloud-enable this application** checkbox

Here is the new **Mobile Project** wizard populated with the information that we covered in the preceding table:

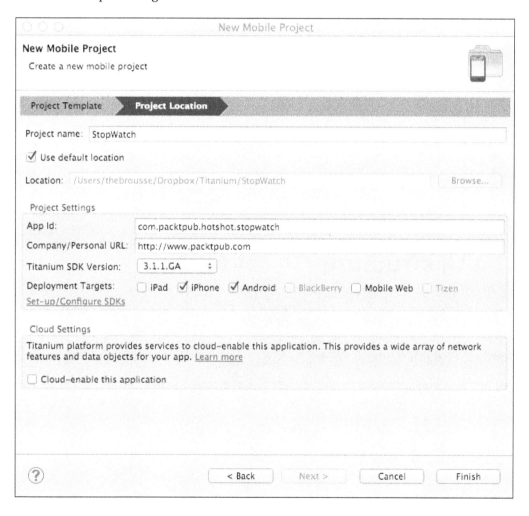

Once you have assigned an App ID to a project, you cannot change it. The reason for this is because, during its creation, Titanium generates a **guID** (a technical key if you will). So if you were to change an application's App ID, it wouldn't match the guID anymore. Therefore, it is recommended that if you ever needed to change an App ID, simply create a new project and move the source files into this newly created project.

What have we here?

The wizard created several files and directories. We will cover what those files are as we go forward, but for now let's just turn our attention to the ones that we will focus on in this chapter:

- `tiapp.xml`: It contains the application's metadata. This ranges from description, technical IDs, and custom settings, to name a few.

- `app.js`: It contains the source code for our application.

Now technically, this is already an application. If we were to run the generated code, we would have a fully functioning application. Of course, at this point in time, it is pretty rudimentary.

The UI structure

Before we start placing UI elements on the screen, we need to have a coherent vision of how the controls will be placed on the screen. Since Titanium Studio does not provide us with a visual editor (as of writing this book), we need to determine where the controls will be placed (x and y coordinates) and how big they will be (width and height).

Our application's user interface will comprise of a single window and will be divided into three sections (views). Inside these views, we will place controls such as labels, buttons, and a scrolling list view.

The following figure is a visual representation of how the user interface elements will be stacked atop one another:

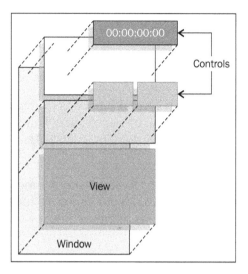

Why do we use views?

Views act as containers; they are usually used to group controls together and can be more easily moved around the screen if needed. One very simple example for this would be a toolbar; it is nothing else than a panel containing buttons when you think of it. Now, if you need to change the location of the toolbar on your screen, you can simply move the container, not every single button. A window is a top-level container that can contain other views. The major difference is that it can be opened and closed. Opening a window will load all of its containing views; closing that same window will automatically remove the views contained in it.

> For developers coming from the Java or .NET world, views would be equivalent to `Panels`. As for developers more familiar with HTML, they can be considered as `DIV` tags.

Now on to the code

As mentioned earlier in this chapter, all of the source code for the application is located in a single file (`app.js`). For every chapter in this book, we will delete all of the code contained in this file and replace it with our own. We will iterate on it until we have a complete working application.

It all starts with a window

Now that we have opened the `app.js` file and cleared all of its content, we can start working on a clean slate.

So the very first thing that we need in every application is a `Window` object. Every application today has at least one window. Some of them can fill the entire screen, have no title whatsoever, or even be transparent. But they are windows nonetheless.

The Titanium API provides us with several functions to create our UI objects. They are usually contained in the `Titanium.UI` namespace. Think of it as a package where you store functions that share a similar domain (such as User Interface, Geolocation, and Media management).

We create the window using the `Ti.Ui.createWindow` function. We want it to have a white background color and a vertical layout. Since we want to interact with our window later on in the code, we store its reference into a variable named `win`.

```
var win = Ti.UI.createWindow({
  backgroundColor: '#ffffff',
  layout: 'vertical'
});
```

 Some of you may have noticed that we used the `Ti.UI` prefix instead of `Titanium.UI`. This is just a shortcut in order to make the code more readable. Both forms will work and behave in the same manner when building the application.

We added the `layout` property as an extra property to make sure that components will be stacked below one another. Thus, we're making sure that none of them will overlap. Different applications will have different needs, but in this specific case, it is appropriate.

Displaying the current time with big numbers

The main function of a stopwatch is to display the current time; for this, we will use a `Label` component. In order to give a visual effect of having light text over a dark background, we will put this label in a container view as explained in *The UI structure* section of this chapter.

So first, we create our container view. We want to position it at the very top. It will fill the entire width of the screen and will take up 30 percent of the screen's height.

```
var timeView = Ti.UI.createView({
  top:0,
  width: '100%',
  height: '30%',
  backgroundColor: '#1C1C1C'
});
```

Then, we want to create a label that will display the timer itself. The label object has more than 50 properties that can affect its appearance and behavior, so we won't go over all of them at this stage. What is important to remember here, is that we want it to have a large font (to display big numbers), and we want the text to be centered. We will also give a default value of READY? for when the counter is not running.

```
var label = Ti.UI.createLabel({
  color: '#404040',
  text: 'READY?',
  height: Ti.UI.SIZE,
  textAlign: 'center',
  verticalAlign: Ti.UI.TEXT_VERTICAL_ALIGNMENT_CENTER,
  font:{
    fontSize: '55sp',
    fontWeight: 'bold'
  }
});
```

Instead of specifying the height of the label, we used the Ti.UI.SIZE constant. This means that the control will adapt its size automatically to fit its content. This is particularly useful when we can't predict how much content will have to be displayed at runtime.

Now that both controls have been created, we need to add them to our window. As explained earlier, first we add the Label object to its container view:

```
timeView.add(label);
```

Then, we add this same container view to our window:

```
win.add(timeView);
```

Finally, we show the window using the open function:

```
win.open();
```

We can now do our first run

We can launch the application using the **Run** button from the **App Explorer** tab (on a simulator/emulator, or directly on a device). To see your code in action, simply click on the **Run** button, select where you want to run it, and the application will start (provided the code has no errors).

Our very first run shows our main window (with a white background), our container view (with a dark background), and our label showing **READY?** on the screen, as shown in the following screenshot:

This is not much yet, but we are getting there.

Starting and stopping the stopwatch

Now that we can show the time on our stopwatch, we need some way to start and stop it. To do this, we will use two button components. Buttons are pretty straightforward; the user touches it, an event occurs, and an action is performed.

As we want our buttons to be horizontally aligned, it makes perfect sense to group them inside a view. This view will act as a toolbar. It will fill the entire width of the screen, occupy 10 percent of the screen's height, and have a horizontal layout. The buttons are laid out from left to right.

```
var buttonsView = Ti.UI.createView({
  width: '100%',
  height: '10%',
  layout: 'horizontal'
});
```

Now, we need a button to start the stopwatch and another one to stop it. Much like labels, buttons have a lot of properties, so we won't go over all of them here. What is important here, is that they each have a different title (this is what the user sees on the button), and they each take up about 50 percent of the screen's width. Also, they each have different colors but share the same font.

```
var buttonStartLap = Ti.UI.createButton({
  title: 'GO!',
  color: '#C0BFBF',
  width: '50%',
  height: Ti.UI.FILL,
  backgroundColor: '#727F7F',
  font: {
    fontSize: '25sp',
    fontWeight: 'bold'
  }
});
```

 Notice that we defined the font size in sp units — also known as **Scale-independent Pixels**; an abstract unit that is based on the physical density of the screen. It is recommended to use this unit when specifying font sizes so they will be adjusted for both the screen density and the user's preference.

```
var buttonStopReset = Ti.UI.createButton({
  title: 'STOP',
  color: '#C0BFBF',
  width: '50%',
  height: Ti.UI.FILL,
  backgroundColor: '#404040',
  font: {
    fontSize: '25sp',
    fontWeight: 'bold'
  }
});
```

 Contrary to our timer label, the height property uses the Ti.UI. FILL constant, which means the buttons will grow to fill their parent's height (the toolbar view's height).

Notice that we didn't give any x position for the buttons. That's one advantage of using the layout property since it takes care of it for us.

Once the buttons are created, we simply add them to their parent view and then add this same parent to the main window, as shown in the following code:

```
buttonsView.add(buttonStopReset);
buttonsView.add(buttonStartLap);
win.add(buttonsView);
```

We see the buttons, but they don't do much yet!

Once we have added our buttons, we can run our application once again, and we will see our newly added buttons. But at this point in time, they have no code behind them, as shown in the following screenshot:

Let's start with the **GO!** button. We want it to start counting when the user presses it. So to achieve this, we need to add an event handler that will be triggered when the user presses the button.

Titanium provides us with a simple function named addEventListener. This function has two parameters:

- The event's name we want to handle (click, swipe, pinch, and so on)
- The function that will be executed when the event occurs

For our two buttons, we add event handlers for the click event. Once this event is caught, the code contained in the function (passed as a second parameter), will be executed, as shown in the following code:

```
ButtonStartLap.addEventListener('click', function(e) {
  stopWatch.start();
}

ButtonStopReset.addEventListener('click', function(e) {
  stopWatch.stop();
  label.text = 'READY?';
}
```

We could have also used an already existing function and reused it as our second parameter. This becomes useful when you want to reuse the same code for different events.

```
var alertFunc = function(e) {
    alert('Reusable code');
}

buttonStartLap.addEventListener('click', alertFunc);
```

Stopwatch is not defined, but it's okay

If you have tried running your code after adding event handlers, you will be shown an error message related to the fact that the stopWatch object is undefined. That's because there is no such out of the box feature in Titanium. Therefore, we will have to create our own.

It's quite simple really; it has four functions in total, as shown in the following table:

Function name	What it does
start()	Starts the watch.
stop()	Stops the watch.
toString()	Returns a human-readable version of the elapsed time (in the 00:00:00:00 format). It has to work even while running.
reset()	Resets the watch to 0.

It also has to provide a listening mechanism in order to be able to perform an action at a specific interval (every 10 milliseconds in our case).

We are not going to cover the code from these functions in this book, since there are already plenty of JavaScript examples available on the web or the Internet. Also, putting the code used to calculate the difference between two timestamps would prove pretty much impossible to read on paper.

You can develop your own implementation, and as long as your code provides the methods needed, it will integrate seamlessly into the existing code.

But we took the liberty of providing one in the application's code, which is available on the public GitHub repository at `https://github.com/TheBrousse/TitaniumMobileHotshot`.

To use the library provided with the sample, we have to do the following:

1. Load the stopwatch functionality contained in the `stopwatch.js` file located next to our `app.js` file. We can do this using the `require` function that follows the `CommonJS` pattern. There is a lot of documentation available regarding this pattern, and we will cover this topic in a chapter as we move forward. But for now, just see it as a way to load a feature that is contained in another file.

   ```
   var Stopwatch = require('stopwatch');
   ```

2. We then need to create a function that will be triggered every time the stopwatch reaches a fixed duration in milliseconds (10 should allow us enough precision, without having a huge impact on performance). It contains only one statement that updates the value of the label with the current time.

   ```
   function stopwatchListener(watch) {
     label.text = watch.toString();
   }
   ```

Since it is called repeatedly, it is recommended that such a function performs only small operations to avoid degrading the performance.

3. Then, we create a stopwatch object by specifying the listener function that will be attached to it (`stopwatchListener`) and the interval between calls (10 milliseconds), as shown in the following code:

   ```
   var stopWatch = new Stopwatch(stopwatchListener, 10);
   ```

With this code in place, we already have a fully-functioning stopwatch application that we can use. If we press the **GO!** button, the timer will start. If we press the **STOP** button, it will stop. And if we press the **Go!** button again, the timer will continue from where it was when it stopped.

Keeping up with lap times

Since we now have a working stopwatch application, we can implement more advanced features. One of those new features is the ability to keep a track of each lap and later on being able to view those laps.

The way that this feature will be presented to a user is quite simple. When the timer is running, the user has the ability to press a button. When this button is pressed, the application saves the current value on the timer and adds it to a list, without affecting the stopwatch counter in progress. The user can use this feature for an unlimited number of times, as long as the stopwatch is running.

Capturing lap times

Our application has only two buttons, and in the interest of good design, we will reuse those buttons so that they can offer other features when not in use. Therefore, the **GO!** button will become the **LAP!** button once the counter is started.

To keep track of whether the timer is already running or not, we will simply use a `Boolean` variable:

```
var isRunning = false;
```

We must modify our event listener for the **GO!** button to reflect what we want to achieve. We first check whether the timer is already running. If that is not the case, we change the variable's value so we know it is now running. We then change the button's title and start the timer, as shown in the following code:

```
buttonStartLap.addEventListener('click', function(e) {
  // If the timer is running, we add a new lap
  if (isRunning) {
    Ti.API.info(stopWatch.toString());
  } else {
    // If the clock is not ticking, then we start it
    isRunning = true;
    buttonStartLap.title = 'LAP!';
    buttonStopReset.title = 'STOP';
    stopWatch.start();
  }
});
```

If it is already running, it means that we want to save a lap time. In this instance, we just log its value to the console using the `Ti.API.info` logging function.

Titanium's logging API is a very useful feature that allows displaying messages to the console without having to resort to alert dialogs or other obscure mechanisms. It also allows us to assign different severity levels depending on the messages we want to log (info, warn, error, debug, and trace).

After the first run and a few laps, we will have an output to the console that will look like this:

```
[INFO]  00:01:17:37
[INFO]  00:02:53:19
[INFO]  00:04:06:93
[INFO]  00:06:11:52
```

This is not much to show for now, but the feature is there nonetheless.

Showing lap times in a scrollable list

Now that we are able to gather lap times and log them to the console, we want to display those values in a scrollable list. This new `TableView` component will have a light background color. It will fill the entire width of the screen and will occupy all of the remaining available height of the screen.

```
var table = Ti.UI.createTableView({
  width: '100%',
  height:Ti.UI.FILL,
  backgroundColor: '#C0BFBF'
});
```

Once the table view is created, we add it to the main window.

```
win.add(table);
```

Now that we have our list on the screen, we need to replace the code section where we logged the time to the console earlier. Instead, we will create a `TableViewRow` object. Much like labels, they have many properties to customize their appearance and behavior. But those that deserve attention here, are:

- `title`: It is the text displayed on the row.
- `leftImage`: It is the image that will be displayed in front of the row text. Here, the image is in the `Resources/images` directory, but it could be located anywhere as long as it is somewhere under the `Resources` directory.
- `className`: It is the unique row layout. It can be any string you want, as long as all of the rows that share the same layout have the same `className` property.

Setting the `className` property is very important since it is used to optimize rendering performance. It enables the operating system to re-use table rows that are scrolled out of view to speed up the rendering of newly visible rows. So, always check that you assign this property if you encounter performance issues with table views (especially with a lot of rows).

Other properties are used for cosmetic purposes (in the context of this application).

```
buttonStartLap.addEventListener('click', function(e) {
  if (isRunning) {
    var row = Ti.UI.createTableViewRow({
      title: stopWatch.toString(),
      color: '#404040',
      className: 'lap',
      leftImage: '/images/lap.png',
      font:{
        fontSize: '24sp',
        fontWeight: 'bold'
      }
    });
```

Finally, we want to add this newly created row to the table using the `appendRow` function.

```
    table.appendRow(row);
  } else {
  // else code doesn't change
```

Resetting the timer

The **reset** button shares the same principle as the **GO!** button. When the timer is running, the button gives the ability to stop the timer. If it is already stopped, it can be used to reset the timer and empty the list of laps.

We check whether the timer is already running. If it is, we stop the timer, change the button title accordingly, and set the variable to `false`.

If it is not already running, we clear the table of all its rows. Notice that we don't really delete any row, but instead we assign its data with an empty array. Then, we reset the timer and update the label's text to its initial value.

```
buttonStopReset.addEventListener('click', function(e) {
    if (isRunning) {
      buttonStartLap.title = 'GO!';
      buttonStopReset.title = 'RESET';
      stopWatch.stop();
      isRunning = false;
    } else {
      table.setData([]);
      stopWatch.reset();
      label.text = 'READY?';
    }
});
```

Well, there you have it! A fully-featured, native mobile application that is fully compatible with iOS and Android devices with a single code base. All of this was made with less than 120 lines of code (including comments).

Here is our final stopwatch with the lap counter:

Summary

In this first chapter, we created an entire mobile application from scratch using Titanium Studio. We also saw how important it is to have a clear vision in terms of the user interface; otherwise, things can become complex when you need to move them around.

We also went through the process of creating our user interface using views, labels, buttons, and even a table view. We also learned how to style and place them to meet our needs. We managed to get those same components to interact with one another using events.

While testing and debugging our application, we had a brief introduction to Titanium's logging API.

In the next chapter, we will go over the use of audio files, including reading and writing them to the filesystem and table views.

2
Sili, the assistant that just listens

In this chapter, we will be making our very own voice recorder application called Sili. While other popular applications interpret voice commands, this application simply listens by storing the recordings on the device. We will also provide the possibility of listening to previous recordings by selecting them from a list. We will cover **user interface** (**UI**) interactions of course, but we will also be covering media and filesystem interactions.

By the end of this chapter, you will have learned the following concepts:

- Applying images to standard controls (such as buttons) to implement a "custom" look and feel
- Recording audio and saving to the device
- Interacting with the filesystem
- Playing back an audio file
- Deleting a specific table view row

Titanium only supports audio recording on the iOS platform at this time. Therefore this project will be specifically built for an iPhone. Also, while all the interface interactions can be tested using the simulator, you will need a physical device in order to test the voice recording feature properly.

Creating our project

First, we need to set up a new project for our application. To do this, select the **File | New | Mobile Project** menu from the Titanium studio and fill out the wizard forms with the following information:

Field	Values to be entered
Project Template	Default project
Project Name	**Sili**
Location	You can either:
	• Create the project in your current workspace directory by checking the **Use Default Location** checkbox
	• Create the project in a location of your choice
App Id	`com.packtpub.hotshot.sili`
Company/Personal URL	`http://www.packtpub.com`
Titanium SDK Version	By default, the wizard will select the latest version of the Titanium SDK, which is recommended (as for this writing, we are using Version 3.1.1 GA)
Deployment Targets	Only check **iPhone**
Cloud Settings	Uncheck the **Cloud-enable this application** checkbox

Project creation is covered more extensively in *Chapter 1, Stopwatch (with Lap Counter)*. So, feel free to refer to this section if you want more information regarding project creation.

The user interface structure

This new application's user interface will also be comprised of a single window, divided into sections using views as containers:

- The top view will act as a header containing the application's title using a **Label** and a **Button** for interaction.
- The second view will be the largest one of them all. Spanning most of the screen's height, it will contain a list of all the saved recordings.

- The last section will be present at the bottom of the application and will contain a very large **Button** so that it is easily accessible, as shown in the following figure:

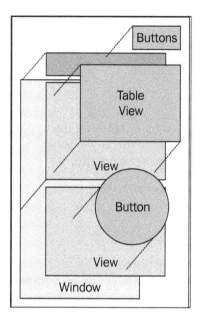

Coding the application

Now that we have a better idea of how our application will look, and most importantly, how it will be structured, we will go ahead and translate that into code.

Let's do some scaffolding

We can start creating our window and views, and add controls to them. While these controls won't have any interaction (yet), it will be much easier down the road to implement interactions once everything is in place. As we did before, we will open the app.js file and clear all its content in order to have a clean slate.

Since every single application needs at least one window, ours is no exception to the rule. So, we will create one using the `Ti.UI.createWindow` function and set its `title` to `Sili`. We also need to add additional views and controls later on in the code, hence we store its reference in the `win` variable:

```
var win = Ti.UI.createWindow({
  backgroundColor: '#ffffff',
  title: 'Sili',
  layout: 'vertical'
});
```

Our top view will serve as a header containing the application's title as well as two buttons. We want it to have a dark blue background. It will fill the entire width of the screen and will span `10%` of the screen's height:

```
var headerView = Ti.UI.createView({
  height: '10%',
  width: '100%',
  backgroundColor: '#002EB8'
});
```

We also need to create a label, whose text will be white and will be vertically aligned with a more prominent font than conventional labels. It will also use the window's title so that they both share exactly the same value:

```
headerView.add(Ti.UI.createLabel({
  text: win.title,
  left: 7,
  color: '#ffffff',
  height: Ti.UI.FILL,
  verticalAlign: Ti.UI.TEXT_VERTICAL_ALIGNMENT_CENTER,
  font:{
    fontSize: '22sp',
    fontWeight: 'bold'
  }
}));
```

You may have noticed that the label's reference was not stored in a variable. This is a shortcut you can use when you need to add a control to a container and you know you won't be accessing it down the line. This makes for a linear code and removes the overhead of creating a new variable on the device during execution.

The header view also contains two buttons for interaction. One to activate the list's edit mode and another one to deactivate it. We'll get to that later, but for now, let's create and add them to the view.

The important thing to notice here is that both of the buttons will be located at exactly at the same location on the screen. This will give the effect of using a single button to toggle the list's edit mode. While in fact, the user will be interacting with two different buttons that are shown (or hidden), depending on the context. Therefore, one button will have the `visible` property set to `false` and the other one set to `true`:

```
var edit = Titanium.UI.createButton({
  title: 'Edit',
  right: 5,
  visible: true
});
var done = Titanium.UI.createButton({
  title: 'Done',
  right: 5,
  visible: false
});
```

We then add each button to the container view:

```
headerView.add(edit);
headerView.add(done);
```

Finally, we add the `headerView` object to the main window:

```
win.add(headerView);
```

The second container view will contain a list of all the audio recordings saved by the user. It will occupy 65% of the screen's height and 100% width:

```
var recordingsView = Ti.UI.createView({
  height: '65%',
  width: '100%'
});
```

The `TableView` component is best suited for displaying items in the form of lists. Since it will be contained in the view created previously (`recordingsView`), we simply use the `Ti.UI.FILL` constant for setting the `TableView` component's `width` and `height` properties. With this, the component will grow accordingly in order to occupy its parent's entire surface. We also make it editable in order to allow the users to modify the table's content:

```
var table = Ti.UI.createTableView({
  width: Ti.UI.FILL,
  height: Ti.UI.FILL,
  editable: true
});
```

We will then add the table to its parent view and the newly created view to the main window using the following code:

```
recordingsView.add(table);
win.add(recordingsView);
```

Our last view will be placed at the bottom of the screen and will contain a single, large button. It will have a dark background, and will span `25%` of the screen's height, and `100%` of the width:

```
var buttonView = Ti.UI.createView({
  width: '100%',
  height: '25%',
  backgroundColor: '#404040'
});
```

Now, if you remember the user interface structure schematics, you may have noticed a round button. Since such a look and feel is not achievable using regular buttons, we shall rely on an `ImageView` object and use it just as we would use a regular button. We simply assign it an image and set its `height` property to `95%` to provide some padding, as shown in the following code:

```
var recordButton = Ti.UI.createImageView({
  image: '/images/recording_off.png',
  height: '95%'
});
```

 We haven't set any height to our `ImageView` component, since the image file has the same width and height (a perfect square). Therefore, the `ImageView` component will automatically scale while keeping the image proportions intact.

As before, we will add the image view to its parent view and then add the said parent view to the main window, using the following code:

```
buttonView.add(recordButton);
win.add(buttonView);
```

In order to actually see something on the screen, we need to display the window:

```
win.open();
```

Let's see how this looks

Now that all user interface elements are in place, we are ready for our first run. For this, all we need to do is just click on the **Run** button from the **App Explorer** tab. It was mentioned at the beginning of this chapter that you need an actual device to run this application. At this stage, there is no specific code that would prevent you from running it using the simulator.

The following screenshot appears on our first run:

Of course, these controls don't respond to any user input and the list is empty. But this now gives us the structure in which we need to include the core logic of the application.

Recording with a click on a single button

Now let's dive into our application's main feature, the audio recording. Titanium provides us with an `AudioRecorder` object that can be used to record from the device's microphone.

It goes pretty much similar to the following steps:

1. We create an `AudioRecorder` object.
2. We start the recording using the `start` function.
3. We stop the recording using the `stop` function.
4. This function will return a `File` object, which we can interact with and save on the device's filesystem.

The first thing we need to do is to set the audio session mode. There are a few modes to choose from, depending on the type of recordings you want to do. In our specific case, we want to be able to record and playback, but not simultaneously. This can be done using the following code:

```
Ti.Media.audioSessionMode = Ti.Media.AUDIO_SESSION_MODE_PLAY_AND_
RECORD;
```

Next, we need to create our `AudioRecorder` object and store its reference in a variable named `recorder`, in order to interact with this object in the code:

```
var recorder = Ti.Media.createAudioRecorder();
```

We also need to set the audio properties of our recorder. Again, there are many options and combinations to choose from. But in our case, we will use the 8-bit ULAW encoding format, which is quite appropriate for voice recording. As for the file format, we will use WAVE files. It could have been MP3 or another format, but WAVE files will be easier to use later on for playback:

```
recorder.compression = Ti.Media.AUDIO_FORMAT_ULAW;
recorder.format = Ti.Media.AUDIO_FILEFORMAT_WAVE;
```

Now that we have our audio recorder in place, let's add some behavior to our big **Record** button using the `click` event handler. Since we are using only one button to start and to stop recording, we need to have a different behavior depending on whether we are recording or not. To achieve this, we can rely on the `recording` property of the recorder. If we are recording, then we stop; if we are not, simply start recording. We also need to change the button's image in order to give a visual indication to the user that the application is actually recording:

```
recordButton.addEventListener('click', function(e) {
  if (recorder.recording) {
    var buffer = recorder.stop();
    e.source.image = '/images/recording_off.png';
  } else {
    recorder.start();
    e.source.image = '/images/recording_on.png';
  }
});
```

To change the button's image, we can just use the `recordButton` variable. But another way of doing it is through the source of the event. Most events pass a variable to the event function (e in our case). Logically, `e.source` should return the button itself. This is very useful while working dynamically with created controls, and we don't know which specific button has called the event handler. This is usually called the **DRY** (**Don't repeat yourself**) principle, meaning that there should only be one copy of any non-trivial piece of code. This is considered good practice.

What good is a buffer if we're not using it?

For now, when the recorder stops, the return value is stored in a variable named `buffer`. But since we don't do anything with it after that, the recording is lost. Let's remedy that by saving the buffer to the devices filesystem.

First, we must determine where the recordings will be stored. Titanium provides access to the application's data directory using a simple property (`Ti.Filesystem.applicationDataDirectory`). We generally use this directory to store application-specific files, which is the perfect location to store our recordings.

Let's create a variable that will act as a shortcut for the application data directory, thus making the code more concise:

```
var APP_DATA_DIR = Ti.Filesystem.applicationDataDirectory;
```

We need to retrieve the recording's content right when the user stops recording and write it onto the filesystem. To do this, we create a new empty file using the getFile function with the location and the filename as parameters. In order for each file to have a unique name, we use the getTime function of the Date object (which returns the number of milliseconds since 1970/01/01) and append the .wav file extension.

Once the new file is created, we write the entire content of the buffer onto it using the following code:

```
recordButton.addEventListener('click', function(e) {
  if (recorder.recording) {
    var buffer = recorder.stop();
    var newFile =Titanium.Filesystem.getFile(APP_DATA_DIR,
      new Date().getTime() + '.wav');

    newFile.remoteBackup = true;
    newFile.write(buffer);

    e.source.image = '/images/recording_off.png';
  } else {
// else code doesn't change
```

> Setting the remote backup to true will ensure that the file will be backed up to Apple's Cloud service, meaning that in cases where the user reinitializes his or her device, the recording he or she created will not be lost.

Listing stored recordings

Even though we have the ability to record multiple recordings and save them to the device, there is no way to know for sure that the recordings are indeed on the device. We need to create a function that will read all of the files contained in the application directory, loop through each of them, and then fill the list view.

In the declaration section, we get the reference application data directory and we list all of the files contained in the directory. Since the table view is expecting an array for its content, we create an empty array that will be filled later on in the function:

```
function loadExistingAudioFiles() {
  var dir = Ti.Filesystem.getFile(APP_DATA_DIR);
  var files = dir.getDirectoryListing();
  var tableData = []
```

We now need to loop through the list of files found in the directory using the following code:

```
for (var i = 0; i <files.length; i++) {
```

But since the list only contains the filenames and not the objects themselves, we must first instantiate each file in order to retrieve the metadata, such as the timestamp of the creation, since the only way that we can distinguish one recording from another is by its timestamp. We will then format this information and use it as the label for each row:

```
var recording = Ti.Filesystem.getFile(APP_DATA_DIR, files[i]);
var ts = new Date(recording.createTimestamp());
var rowLabel = String.formatDate(ts, 'medium') + ' - '+ String.
formatTime(ts);
```

We can now create the `TableViewRow` object using the value we formatted previously as its `title` attribute:

```
var row = Ti.UI.createTableViewRow({
  title: rowLabel,
  leftImage: '/images/tape.png',
  color: '#404040',
  className: 'recording',
  font:{
    fontSize: '24sp',
    fontWeight: 'bold'
  },
  fileName: APP_DATA_DIR + '/' + recording.name
});
```

 Pay close attention to the `fileName` attribute. This property is not a part of the Titanium API, but since we are using JavaScript, we have the liberty to add additional properties which we can use later on in our code. This is very useful in a case similar to ours, where we will need extra information when the user selects a row from the list.

Once the row is created, we add it to the data array and once we are done creating all of the `TableViewRow` objects, we set the array to the table view using the following code:

```
tableData.push(row);
} // for end
table.setData(tableData);
} // function end
```

Be sure to call it!

Even if we had already made several recordings, the current code base won't show anything in the list. The reason is quite simple; the following function is called from anywhere in the code:

```
loadExistingAudioFiles();
```

To the following locations:

- Right before the main window opens (this will load the list when the application launches)
- Inside the `recordButton` click event handler, right after the file is written on the device (this will refresh the list every time a new recording is created)

Now that our function is called, we now have a list of all our recordings on the screen, as shown in the following figure:

Listening to a recording

The playback functionality of our application is pretty straightforward. When the user selects an item from the list, it will be played. This is where Titanium shines with simplicity, thanks to the `Titanium.Media.Sound` object that allows audio playback with very little code.

First, we need to add an event listener for the `click` event on the table view. This will give us access to the selected row through the `e.rowData` variable. Since this is the same as the table view row we created earlier, it has the `fileName` custom property that we assigned to it. It represents the file's complete path and will be used as the `url` property to create the sound object. We then start the playback using the following code to have a pretty functional voice recorder:

```
table.addEventListener('click', function(e) {
  var sound = Ti.Media.createSound({
    url: e.rowData.fileName
  });
  sound.play();
});
```

Deleting old recordings

The last feature missing from our application is the ability to delete recordings. To achieve this, we will be using the two buttons from the header view (one visible and one invisible at any given time) as well as the table view's built-in editing functionality.

The user interactions will be as follows:

1. When the user switches onto the edit mode by clicking on the **Edit** button, the following events will happen:

    ○ The table view's editing mode is switched on

    ○ The **Edit** button turns invisible

    ○ The **Done** button turns visible

2. From there, the user can select rows for deletion and confirm by clicking on the **Delete** button generated by the table view.

3. When the user comes out of the edit mode by clicking on the **Done** button, the following events will happen:

    ○ The table view's editing mode is switched off

    ○ The **Edit** button turns back visible

    ○ The **Done** button turns invisible

Can you see a pattern here? No matter which button the user clicks on, it does the complete opposite of each operation. In the cases similar to this, it is easier to encapsulate
these changes into a single function that sets each value to its opposite using the following code:

```
function toggleEditMode() {
  edit.visible= !edit.visible;
  done.visible= !done.visible;
  table.editing= !table.editing;
}
```

Once the logic is coded in a single function, both buttons call it and will have the desired effect:

```
edit.addEventListener('click', function(e) {
  toggleEditMode();
});
done.addEventListener('click', function() {
  toggleEditMode();
});
```

If we now run the application, we will observe the desired effect. The table view's edit mode allows us to delete rows from the list, and the two buttons will act as a toggle between the two modes, as shown in the following screenshot:

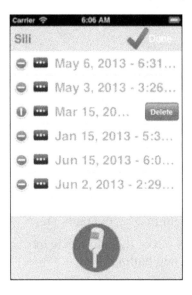

Now, if you look a little more closely, you will notice that we are not fully there yet. While the rows are deleted from the list, it is not (yet) the case for the actual recordings. In fact, if you were to delete a few rows, shut down the application, and restart it, you will notice that all the recordings are back.

In order for changes on the table to be reflected on the filesystem, we need to add an event listener to the table view for the `delete` event. Inside this listener, we will retrieve the complete file path from the row and instantiate a file object matching the said path. We will then delete the file using the following code:

```
table.addEventListener('delete', function(e) {
  var fileToDelete = Ti.Filesystem.getFile(e.rowData.fileName);

  fileToDelete.deleteFile();
});
```

 You don't need to delete the row from the table view since this is done automatically by the component. The event listener serves as an "extra action" that you would want to do when the visual row is deleted.

This is it for our second mobile application. A complete audio-recorder that allows you to record audio, save them on your device, play them back, and delete them when needed. You can, of course, imagine a lot of new features to add based on the code you already have, such as volume control, audio quality, files format, and many others. Feel free to experiment with what the framework can offer and see the endless possibilities.

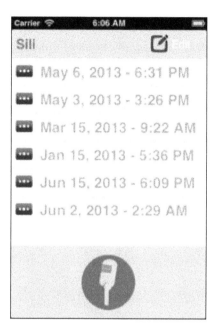

Summary

In this chapter, we learned how to use the audio capabilities of Titanium. We also learned how to interact with the device's filesystem and how to read and write from the application data directory. Finally, we learned how to use the powerful table view's edit mode.

In the next chapter, we will learn how to save data on our device using a database.

3
The To-do List

In this chapter, we will be building a simple, yet useful application to manage To-do items. This application will allow us to create tasks, mark them as completed, and will also provide us with the possibility to delete all the completed items by pressing a single button.

Another aspect covered in this chapter is the use of the SQLite-embedded database. Persisting all our items in such a database will allow our tasks to exist on the device even after the application is closed.

By the end of this chapter, you will have learned the following concepts:

- Creating a database structure
- Adding records to a database
- Listing all the items from a database
- Updating an existing item from a database
- Navigating through the data structure of a `TableView` component
- Deleting items from a database based on a filter

Creating our project

As with our previous projects, we need to set up a new project for our application. To do this, navigate to **File | New | Mobile Project** from Titanium Studio, and fill out the wizard forms with the following information:

Field	Value to enter
Project Template	**Default Project**
Project name	`To Do List`
Location	You can either:
	• Create the project in your current workspace directory by checking the **Use default location** checkbox
	• Create the project at a location of your choice
App Id	`com.packtpub.hotshot.todo`
Company/Personal URL	`http://www.packtpub.com`
Titanium SDK Version	By default, the wizard will select the latest version of the Titanium SDK. This is recommended (as of this writing, we are selecting Version **3.1.3 GA** from the dropdown).
Deployment Targets	Check the **iPhone** and **Android** options.
Cloud Settings	Uncheck the **Cloud-enable this application** checkbox.

Project creation is covered in more extensive detail in *Chapter 1, Stopwatch (with Lap Counter)*, so feel free to refer to this section if you want more information regarding project creation.

The user interface structure

Faithful to our simplified design, the application's user interface will comprise of a single window, which is divided into the following three sections using views as containers:

- The top view will act as a header containing a text field and a button used to add a new task to the list.

- The second view will be the largest one of them all. Spanning most of the screen's height, it will contain a list of all the tasks from the database (filtered or unfiltered).

- The last section will sit at the bottom of the screen and will act as a toolbar containing two types of controls. One to toggle between the list's display modes, and a button to delete all the completed tasks.

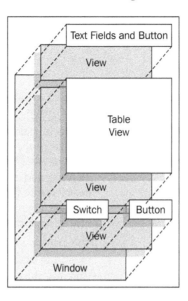

Let's get this show on the road

We will now go through all the steps required to build our application.

The database

Before we dive into the application's look and feel, we need to define how our application will store the information. Titanium provides us with an embedded **relational database management system (RDMS)** called **SQLite**. It is a widely used open source database, which has very low memory consumption, making it the natural choice for mobile development where resources are limited.

Like most relational databases, the information is stored in tables and is accessed using the SQL query language. There are plenty of books and online articles covering all the intricacies of relational databases and SQL. So, we will not go into much detail on this subject, but will cover the basics that are necessary to store, update, and delete information from a database.

Defining the structure

Every time we need to store information in a database, we need to define a structure. For our application, the structure is pretty straightforward. Each task needs a name and some sort of a **flag** defining whether the task is complete or not. So, in our case, a single table would perfectly meet our needs. We shall name it TODO_ITEMS to keep things meaningful.

Our table's structure is as follows:

Column Name	Data Type	Description
ID	INTEGER	The technical unique identifier will allow us to retrieve the task later on, when querying the database. We want it to be automatically incremented every time a new record is inserted into the database. This field's purpose is purely technical and will not be visible from the user's standpoint.
NAME	TEXT	The task's text (or label if you will) that will be visible to the user.
IS_COMPLETE	INTEGER	Whether the task is complete or not, notice that the type is INTEGER while a simple Boolean (Yes/No) would have sufficed. This is due to the fact that SQLite does not support Boolean values.
		To address that, we will simply store 0 or 1 in this column and map those values to their Boolean counterparts in the code.

 For those of you who are unfamiliar with database tables, think of them as spreadsheets, where each column maps to a task attribute, and where each row is a task.

Implementing our model

Now that we have determined the structure of our database, we can now implement the said structure into our code. All database operations are contained in the Titanium.Database namespace and provide a lot of functionalities that will be covered throughout this chapter.

After opening the `app.js` file and deleting all of its content, the next thing we want to do is to open (or create if it is still not present on the device) our database file using the `open` function with the filename as a parameter. This will return a reference to the opened database. If the database does not exist, it will create an empty database file and return a reference to this opened database. We will then store its reference in a variable named `db` for later use in our code as follows:

```
var db = Ti.Database.open('todo.sqlite');
```

Once we have access to our database, we want to make sure that our database structure is present. We then execute the following SQL statement to create the `TODO_ITEMS` table if it is not already present:

```
db.execute('CREATE TABLE IF NOT EXISTS TODO_ITEMS (ID INTEGER
    PRIMARY KEY AUTOINCREMENT, NAME TEXT, IS_COMPLETE INTEGER)');
```

The preceding statement will not raise any error or exception since it contains the `IF NOT EXIST` clause. So there is no extra code needed; by doing this right at the beginning of our code (even before declaring the UI), we are sure that the structure is well in place right from the start.

 While the table creation statement could work without the `IF NOT EXIST` clause, there will be no protection if the statement fails, resulting in data loss.

The user interface

Having previously defined our user interface, we are now able to implement the necessary components of our application. While not yet interactive, it will provide us with a solid baseline to build upon.

As with every single-window application, we create a standard window using `Ti.UI.createWindow` with a white background and `To Do List` as its title. Since we want to add additional controls to this newly created window, we store its reference in the `win` variable as follows:

```
var win = Ti.UI.createWindow({
  backgroundColor: '#ffffff',
  title: 'To Do List'
});
```

The header view

Our header view is basically a container for our controls that will sit on top of the screen. It will have a height of 50dp, spanning the whole width of the screen and will have a blue background. It will also have a horizontal layout, since we do not know how large the screens of some devices are:

```
var headerView = Ti.UI.createView({
  top: 0
  height: '50dp',
  width: '100%',
  backgroundColor: '#447294',
  layout: 'horizontal'
});
```

The header view contains two controls. A text field that is used to enter the task's name and a button to add a new task to the database. The text field will span 75 percent of the screen's width and will display a hint text when the field is empty, providing a better user experience:

```
var txtTaskName = Ti.UI.createTextField({
  left: 15,
  width: '75%',
  height: Ti.UI.FILL,
  hintText: 'Enter New Task Name',
  borderColor: '#000000',
  backgroundColor: '#ffffff',
 borderStyle: Ti.UI.INPUT_BORDERSTYLE_ROUNDED
});
headerView.add(txtTaskName);
```

The button, while square dimensioned, will use the backgroundImage property in order to override the button's default rendering as follows:

```
var btnAdd = Ti.UI.createButton({
  backgroundImage: 'add_button.png',
  left: 15,
  height: '45dp',
  width: '45dp'
});
headerView.add(btnAdd);
```

We will then add the header to our application window:

```
win.add(headerView);
```

The tasklist

The tasklist will use the `TableView` component to display the items contained in the database. As for the header, we will use a `View` container. Its top position will be at `50dp`, which is precisely the height of the header view. So if we were to change the header's height, we would have to update this value as well:

```
var taskView = Ti.UI.createView({
  top: '50dp',
  width: '100%',
  backgroundColor: '#eeeeee'
});
```

 Notice that we did not specify any height for this view. The reason being we already specified the height of the header view (and the footer view).Therefore, the middle view will stretch in order to occupy all the remaining space on the screen.

The table view is created using a limited set of properties since there is no data (yet) to assign to it. We want it to grow in order to fill its parent view; we achieve this behavior using the `Ti.UI.FILL` constant. We will also change the default `separatorColor` property (the thin line between each row):

```
var taskList = Ti.UI.createTableView({
  width: Ti.UI.FILL,
  height: Ti.UI.FILL,
  separatorColor: '#447294'
});
```

We then add the table view to its container and the container to the main window:

```
taskView.add(taskList);
win.add(taskView);
```

The button bar

A view similar to the header will be used for the button bar, but instead, at the bottom of the screen. It will share most of the same properties as the header view, in terms of dimensions and background color. Of course, since we want it to stay at the bottom of the screen, we will do so by setting the `bottom` property to `0`:

```
var buttonBar = Ti.UI.createView({
  height: '50dp',
  width: '100%',
  backgroundColor: '#447294',
  bottom: 0
});
```

The next control is a **switch** that will allow us to hide tasks that are marked as completed. This can be very useful when the user wants to have a glance at what is left to do. We can create it using the `Ti.UI.createSwitch` function, and we want it to display which filter is applied depending on its status (all the tasks when it is on, and only active tasks when it is off). We also want to set its default value; in our case, we want to show all the tasks by default:

```
var basicSwitch = Ti.UI.createSwitch({
  value: true,
  left: 5,
  titleOn: 'All',
  titleOff: 'Active'
});
```

The second control is a regular button that will be used to clear all the completed tasks from the database:

```
var btnClearComplete = Ti.UI.createButton({
  title: 'Clear Complete',
  right: 5
});
```

We then add both the controls to the button bar, and then add the newly created bar to the main window:

```
buttonBar.add(basicSwitch);
buttonBar.add(btnClearComplete);
win.add(buttonBar);
win.open();
```

Let's have a look

Let's take our newest application for a spin by clicking on the **Run** button from the **App Explorer** tab.

The following screenshot depicts what we see on our first run:

Of course, there are no tasks in the list, and the controls have no code behind them. But this is usually a good time to make that sure everything fits nicely and reacts in the right way in terms of the UI. Does the onscreen keyboard mess up the user experience when it is displayed? What about rotation?

Just keep in mind that, at this stage, it is easier to edit or move around the code when there is little to no impact on the application's logic.

Developing a better switch for iOS

For those of you who have tested the code using the iPhone simulator, you may have noticed there was something odd about the switch component. The `titleOn` and `titleOff` properties don't seem to reflect on the iPhone version. This is because on iOS, a switch is only displayed as an on/off switch without any text associated with it.

In the following figure we can see the difference in the implementation between the two versions:

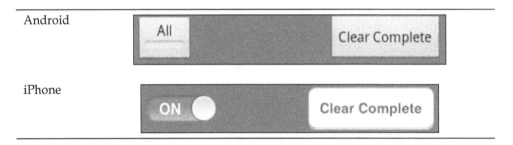

Since the default behaviour on iPhone doesn't suit our needs, we must find a way to get around this problem. Luckily, Titanium allows us to run sections of code, depending on the platform on which the application is installed.

iOS has a component called the **tabbed bar**. It is a button bar that maintains a state by having one button pressed at a time. We can use such a component with two buttons, each having a different title. While the look will be different, the functionality will remain the same. On a positive note, it will have a more native look and feel, since it adheres more to the platform's guidelines.

We must then modify our switch declaration. Instead, we simply declare it and will assign it with proper implementation later. This is made possible thanks to the fact that JavaScript is a dynamic language. We can totally create a variable and assign different values to it, depending on the context:

```
var basicSwitch;
```

If we are running on the iOS platform, we create a new tabbed bar using the `createTabbedBar` function. First, we set its two button labels using an array of string. We want it to have the same background color as the button bar and will give it a special style, which will give it a more compact look. We also want the **All** button to be selected by default, the first element of the `labels` array, by setting the `index` property to zero.

If we are not running on the iOS platform, we use the previously created switch, just as we did in the first version of the code:

```
if (!Ti.Android) {
  basicSwitch = Ti.UI.iOS.createTabbedBar({
    labels: ['All', 'Active'],
    left: 5,
    backgroundColor: buttonBar.backgroundColor,
    style: Ti.UI.iPhone.SystemButtonStyle.BAR,
```

```
      index: 0
    });
  } else {
    basicSwitch = Ti.UI.createSwitch({
      value: true,
      left: 5,
      titleOn: 'All',
      titleOff: 'Active'
    });
  }
```

We could have checked whether we were running iOS by checking whether the Ti.Platform.name property matched the iPhone OS character string. But this would imply an additional verification if we are running on an iPad. Simply checking whether the Ti.Android instance is present is a neat little trick to determine whether we are running Android without performing any string comparison.

We can now see a new control being used when testing on the iPhone simulator. This gives us a look and feel that is more consistent with the platform.

While checking for the platform to run the specific piece of code is useful for small instances such as this one, it can make the code much harder to read and difficult to maintain when faced by large chunks of code that are radically different between platforms.

Titanium addresses this problem by giving us the possibility to load an entire JavaScript file depending on the platform. In the Resources directory, there is a directory for each target platform; we can then use those same directories to provide a different implementation of certain features. As long as the files share the exact same name, Titanium will load the right file.

Here is what it would look like if we were to extract the button bar into its own source file, each having its own implementation. Thus, we would completely avoid the need to check for the platform's name in the code:

What is important to notice here is that both files must have the exact same name (matching the same case).

All platform-specific files (JavaScript, images, media, and so on) are used at build time. The way it works is quite simple. When you build your project, Titanium takes all the files contained in your target directory and copies them into your `Resources` directory. If there is already a file with a similar name, it will overwrite it with the target-specific one. This is a great way to have what you could call a default behaviour for all the platforms, but have a specific implementation for one target for instance.

Adding a new task

With the database set up and all the visuals in place, it is time to add new tasks to our database. We will create a function that inserts a new record into the database using the task's name passed as a parameter. The `execute` function takes two parameters in this case. The first one is the SQL statement, and it is pretty close to English when you look at it closely.

You could read a `SQL INSERT` statement that is pretty much similar to the following:

Insert into the `TODO_ITEMS` table using the following two columns (`NAME` and `IS_COMPLETE`), and the following values: one we don't know yet (hence the question mark), and the second one is zero (for `IS_COMPLETE=false`, since it is a brand new task).

The second parameter is what will replace the question mark in your SQL statement. For every question mark in the SQL statement, there must be a matching parameter in the function call.

Once the task is inserted into the database, we set the text field's value to an empty string to allow the user to type a new one. We also hide the onscreen keyboard if it is displayed using the `blur` function:

```
function addTask(name) {
    db.execute('INSERT INTO TODO_ITEMS (NAME, IS_COMPLETE)
VALUES (?, 0)', name);

    txtTaskName.value = '';
    txtTaskName.blur();
}
```

We call the `addTask` function with the text field's value as a parameter when the user clicks on the **Add** button:

```
btnAdd.addEventListener('click', function(e) {
    addTask(txtTaskName.value);
});
```

We also want to automatically create a new task when the user clicks on the **Return** key on the onscreen keyboard. To do this, we simply fire the `click` event from the **Add** button, which avoids having to copy and paste the code from the `click` event handler:

```
txtTaskName.addEventListener('return', function() {
    btnAdd.fireEvent('click');
});
```

Now, if we create a few tasks using the application, and if we query the database, here is what we would get:

```
sqlite> select * from todo;
    id      name                is_complete
    ----    ----------------    -----------
    1       Dry Cleaner         0
    2       Call Bob            0
    3       Pickup Parcel       0
    4       Buy Milk            0
    5       Walk the dog        0
    6       Balance Checkbook   0
```

This is good because we know our records are inserted correctly.

Listing all the existing tasks

Now that our application can insert tasks successfully into the database, we can now move on to read those same tasks and display them to the user. To do this, we will create a function called `refreshTaskList` that will do just that.

First we need to retrieve all the tasks from the database using a SELECT statement. The `execute` function returns a list of rows (also known as a **ResultSet**) that we will reference as the `rows` variable:

```
function refreshTaskList() {
    var rows = db.execute('SELECT * FROM TODO_ITEMS');
    var data = [];
```

We then need to loop through each row in order to extract the needed information. We can use the `isValidRow` function to determine whether the current row is valid.

Then, we extract the information using the `fieldByName` function and the column's name as a parameter. Since this function can return either a string, a number, or even binary data, we must tailor the data to fit our needs (padding with two empty strings for text on converting 1 and 0 to true or false).

Using the extracted information, we create an object containing all the attributes needed to create a table view row with an extra custom property for the ID. We must not forget to set the `className` property to avoid any drop in performance.

Since the `ResultSet` object is not an array, our code has to move to the next record, or else we would be caught in an infinite loop. Once we pass through all the records, we exit the loop and set the newly created array to the table view:

```
    while (rows.isValidRow()) {
        var isComplete = rows.fieldByName('IS_COMPLETE');

        data.push({
            title: '' + rows.fieldByName('NAME') + '',
            hasCheck: (isComplete===1) ? true : false,
            id: rows.fieldByName('ID'),
            color: '#153450',
            className: 'task'
        });

        rows.next();
    };

    taskList.setData(data);
}
```

Some of you may have noticed the use of the strict equal (===) operator when setting the hasCheck property. This is different from using the equal (==) operator. While they might seem alike and even, sometimes, share the same behaviour, they have one major difference. Equal returns true if the operands are equal. But strict equal returns true if the operands are equal and of the same type. This is very useful in cases where you need to be absolutely sure that the two values are the same.

Before executing the code

While it is true that we have a function that will fill our tasklist, running the code in its present state will not show anything. The reason for that is because the function is not called from anywhere in the code. Therefore, we must add the following line:

```
refreshTaskList();
```

The preceding line must be added to the following locations:

- Right before the main window opens (this will load the list when the application launches)
- At the very end of the addTask function, since we want to refresh the list once a new task is added to the database

Now, if we run the application, we can see all our previously saved tasks displayed in the table view, as shown in the following screenshot:

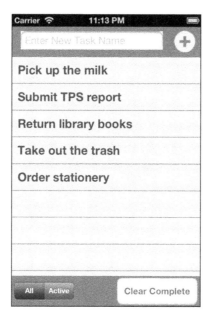

Marking a task as completed

To mark a task as completed, we will need to add an event listener for the `click` event on the table view. From there, we can exactly retrieve which element from the list has been selected, as well as its properties from `e.rowData`.

```
taskList.addEventListener('click', function(e) {
    var todoItem = e.rowData;
```

Then, we determine whether the task is marked as completed using a very straightforward toggle mechanism. If the user selects an incomplete task, we set it as completed. If the task is marked as completed, we simply set its status as incomplete, as shown in the following code:

```
var isComplete = (todoItem.hasCheck ? 0 : 1);
```

We will then update the database record's `IS_COMPLETE` column. To achieve this, we will use the `UPDATE` SQL statement with two parameters (the ID and whether it is completed or not):

```
db.execute('UPDATE TODO_ITEMS SET IS_COMPLETE=? WHERE ID=?',
    isComplete, todoItem.id);
```

Once the database has been updated, we refresh the whole table view:

```
refreshTaskList();
});
```

If we run our application now, we can toggle between tasks depending on their status by clicking on the desired row.

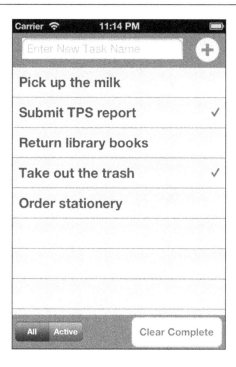

Filtering out completed tasks

When working with large To-do lists, it is often considered useful to show only active tasks. This will make the tasklist more readable and easier to navigate. To do this, we will create a function named `toggleAllTasks` with a Boolean value as a parameter to determine whether we filter the list or not.

If the user chooses to show all the items, we refresh the tasklist. If the user chooses to show only active tasks, we will need to loop through the table view's dataset. For this, it is important to understand that all the **table view rows** are contained in **table view sections**. Even when we don't define any section while populating the table view, there is always one created by default. So, we need to get a reference to this first (invisible) section by accessing the first value contained in the `data` property.

Once we have a hold of the section, we can loop through the `rows` array that contains every row shown on the screen. Looping through each row, we check whether the row is completed or not using the `hasCheck` property. If it is completed, we delete the row by calling the `deleteRow` function by specifying the row index we want to delete:

```
function toggleAllTasks(showAll) {
  if (showAll) {
    refreshTaskList();
  } else {
    var section = taskList.data[0];

    for (var i = 0; i < section.rowCount; i++) {
      var row = section.rows[i];

      if (row.hasCheck) {
        taskList.deleteRow(i);
      }
    }
  }
}
```

Activating the filter

Now that we have our function that can toggle between showing all the tasks and showing only active tasks, we need to link it to our switch (or tabbed bar on the iPhone).

For iPhone, we must add an event listener for the `click` event on the tabbed bar. Inside this listener, we call the `toggleAllTasks` function with the parameter depending on which button was selected. The same event listener must be declared right after the tabbed bar is created.

For Android, we must add an event listener for the `change` event on the switch. Inside this listener, we call the `toggleAllTasks` function with the parameter depending on the switch value. Similar to the tabbed bar listener, this listener must be declared right after the switch is created:

```
if (Ti.Platform.name === 'iPhone OS') {
  // createTabbedBar code unchanged...
  basicSwitch.addEventListener('click', function(e) {
    toggleAllTasks(e.index === 0);
  });
} else {
  // createSwitch code unchanged...
  basicSwitch.addEventListener('change', function(e) {
```

```
    toggleAllTasks(e.value === true);
  });
}
```

Filtering the tasklist to show only the active tasks makes the list easier to read and provides a better user experience, as shown in the following screenshot:

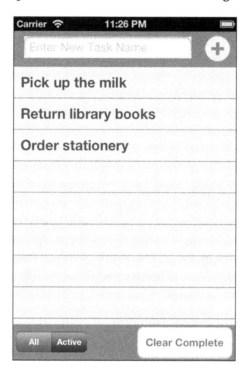

Deleting the completed tasks

Deleting all the completed tasks from the database can be done using a single DELETE statement. Here, the filter is based on records that have the IS_COMPLETE column set to 1 (converted from true earlier).

Once all the completed tasks are deleted from the database, we refresh the table view:

```
btnClearComplete.addEventListener('click', function(e) {
  db.execute('DELETE FROM TODO_ITEMS WHERE IS_COMPLETE = 1;');
  refreshTaskList();
});
```

Close the door, you're letting the heat out

Every time you call the open function on a SQLite database, the system allocates resources into memory in order to perform its operations. In order for these resources to be freed from memory, we must call the close function. Since we open the database when the application launches, it makes perfect sense to close it when the application is closed.

One good way to achieve this is to add an event listener to the main window's close event:

```
win.addEventListener('close', function() {
  db.close();
});
```

Summary

In this chapter, we covered quite a lot of ground. We have learned how to create a database with a defined structure by inserting, updating, and deleting data from it. We also learned how to use new user interface (UI) controls such as text fields, switches, and tabbed bars, and how to use platform-specific code so that the application behaves differently depending on the target. Finally, we learned how to navigate through the table view's data structure, and have a direct impact on which information is presented to the user.

In the next chapter, we will extend Titanium's functionalities using Native Extension Modules.

Interactive E-book for iPad

4

In this chapter, we will be developing an interactive electronic book (e-book) application with realistic page flipping, pretty much like a real book. This will be an iPad-specific application in order to benefit from the large screen resolution offered by a tablet display. We will be using `WebView`, along with HTML and CSS to present a rich experience. We will also be using other rich view components such as `MapView` to show a satellite imagery and a video player.

Since Titanium does not provide "page flipping" functionality out of the box, we will rely on a native module to achieve our goal. We will get more into that later on in this chapter.

By the end of this chapter, you will have learned the following concepts:

- What a native Titanium module is
- Installing, configuring, and using a native module to your existing project
- Using the `WebView` component
- Using the `MapView` component
- Using the `VideoPlayer` component
- Locking the screen's orientation

The user interface structure

Our e-book application comprises a single landscape window, with each page of the book represented as a view that will occupy the entire screen's real estate. All pages will be stacked atop each other and will be shown (or hidden) depending on the current page. We will delegate this page management mechanism to the `Page Flip` Native module, which will do all the heavy lifting for us. Therefore, all our `View` pages will be wrapped in the `PageFlip` component.

Here, the e-book example shown will have only three pages for the sake of this chapter, but you can apply these basic principles on a much bigger scale. The pages are represented as:

- A WebView page referencing a local HTML page
- A MapView page configured to display a specific geographical region
- A VideoPlayer page to play a local video MPEG file

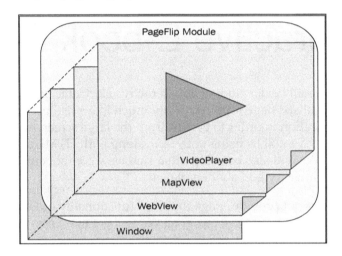

Before writing any code

As mentioned earlier in this chapter, Titanium does not provide any means to give a nice curly effect to the pages while swiping our finger across the screen. There are two ways around this:

- Develop the desired effect using JavaScript by playing with canvas, rotations, and animations. Such an approach would require some fairly complex code and there is also the risk that the performance may be poor depending on the use cases.

- Use a piece of software that uses native code to achieve the desired effect, but still allows us to use it with JavaScript, similar to other Titanium components such as Button, Label, View, and so on. This piece of software is a called a Titanium Native module.

A native module

When using a framework, there will come a time when you will need more than what it has to offer. This may happen progressively as you use the framework more often. The engineers behind Titanium, keeping this in mind, allowed the developers to extend Titanium using native code of their own.

We won't go over the details of developing a Titanium module (we could devote an entire book to this subject). But essentially, it is a piece of code that adheres to a specific set of standards and conventions with additional metadata. Once compiled and packaged properly, Titanium will be able to recognize the module and interact with it using JavaScript, just as if it was any other component of the framework. Modules act as a bridge between your JavaScript code and the platform underneath. Most of the OS functionalities are exposed through the modules.

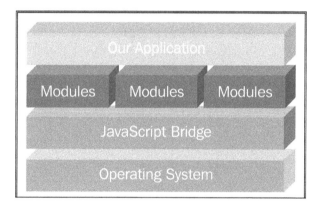

 Not many people know this, but the entire Titanium framework is based on the same principle. Button, Network connections, and Geolocation are all developed as modules under the hood.

Developing a native module will require a pretty good knowledge of the underneath platform. It also should be developed using a specific programming language in order to interact with the native SDK (for example, Objective C for iOS, or Java for Android). So, it is usually destined for more experienced developers. A module is also tied to the platform it is targeting, meaning that if you want a specific feature on different platforms, you will have to implement it on each target platform.

Where do I get it?

Now that we have a better understanding of what a module is and what it can do for us, we will need to actually retrieve it and get it installed on our development machine. There are many ways to retrieve a native module. Some open source modules can be found online, but it usually implies that we need to build them from source.

Another way to get access to the modules is through **Appcelerator Marketplace**. It has about 320 modules at the time of writing this book, with one module being added every second day. For our e-book application, we want to retrieve **Page Flip Module** developed by **Appcelerator** at the following URL: `http://bit.ly/1bykJjd`

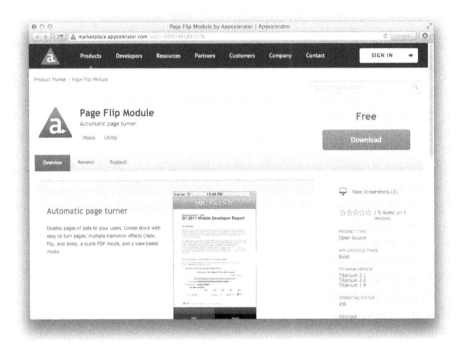

Once logged into the site with your developer account, simply click on the **Download** button.

One important thing to note is that once you have downloaded the module using an account, you will have to use that same account when creating your project later on. Failing to do so will generate an error when running the application thereafter. Also, it is important to create your project after the module has been purchased from the marketplace. This can seem to be a hassle, but it is important to keep in mind, since this is how Appcelerator keeps track of it.

Creating our project

Now that we have our module, we are finally ready to create our project. As with all of our previous projects, we need to set up a new project for our application. To do this, select the **File | New | Mobile Project** menu from **Titanium Studio**, and fill out the **Wizard** forms with the following information:

Field	Value to enter
Project Template	Default Project
Project name	`InteractiveBook`
Location	You can: • Create the project in your current workspace directory by checking the **Use default location** checkbox • Create the project in a location of your choice
App Id	`com.packtpub.hotshot.interactivebook`
Company/Personal URL	`http://www.packtpub.com`
Titanium SDK Version	By default, the wizard will select the latest version of the Titanium SDK. This is recommended. (as of this writing, we are using Version **3.1.3 GA**)
Deployment Targets	Check **iPad** only
Cloud Settings	Uncheck the **Cloud-enable this application** checkbox

Project creation is covered in more detail in *Chapter 1, Stopwatch (with Lap Counter)*; so, refer to this section if you want more information regarding project creation.

Adding the module to our project

Now that we have our newly created project and module downloaded, we can now make the two of them work together.

There are two ways in which we can install the modules, and depending on how you want to use it will determine the best setup that suits your needs. You can either:

- Make the modules available to a single project
- Make the modules available to all projects developed on a given machine

In our case, we want the module to be used with this project only. Therefore, we simply copy the module's archive file (`ti.pageflip-iphone-X.X.X.zip`) to our / `Resources` directory, and then build our project. From there, the Titanium build scripts will extract the module files and copy them to the appropriate directories within the project's hierarchy.

Another option is to extract the files ourselves and copy them at the *root* of our project. This will have the same effect as the previous method, without having to build the project. This is considered the old way of installing modules.

After the module is installed, we need to configure our application in order to actually use it. This is done by editing the `<modules>` XML section in our `tiapp.xml` file. We simply add the following line that identifies the module, its platform, as well as its version:

```
<modules>
    <module platform="iphone" version="1.8.2">ti.pageflip</module>
</modules>
```

Another more user-friendly way of configuring the application would be by using the dedicated graphical user interface from the `tiapp.xml` editor (by selecting the **Overview** tab). From there, we simply have to click on the plus sign (**+**) button and select the `ti.pageflip` module. The editor will then automatically detect the version and the supported platforms from the module's manifest.

If the module installation was done by placing the archive (ZIP) file into the **Resources** directory, and then building the project, there is a very strong chance that the module will already be configured for you. The scripts are becoming more and more efficient with each new release. This is neat since it frees the developers from having "to guess" the required information.

Now we can code

Let's talk about the user interface first.

The user interface

For this application, the user interface is pretty straightforward. As mentioned earlier, we will need to create one view per page, and then delegate the display to the PageFlip component.

Of course, everything has to be contained in a window, so after opening the app. js file and deleting all of its content, we create a standard window using Ti.UI. createWindow with **Interactive eBook for iPad** as its title. We also store its reference in the win variable for later use, as shown in the following code:

```
var win = Titanium.UI.createWindow({
    title: 'Interactive eBook for iPad'
});
```

Importing the PageFlip module

While installing, the module might have seemed complex; using it, on the other hand, is quite simple. Once a module is installed and configured for a project, we can invoke it just as we would invoke any other CommonJS component.

```
var PageFlip = require('ti.pageflip');
```

Thereafter, every time we need to refer to the PageFlip Native module, we can do so through the PageFlip variable.

Adding some colorful views

For the sake of testing the module's behavior, we will create two basic empty views. This will give us the ability to perform the first test very quickly.

```
var orangePage = Ti.UI.createView({ backgroundColor: 'orange' });
var bluePage = Ti.UI.createView({ backgroundColor: 'blue' });
```

We then create our `PageFlip` view that will wrap the two previously created views using the module's `PageFlip.createView` function. We want it to have a curl effect when the user changes the page, and that curl effect should last for `0.3` seconds. We also need to set the `landscapeShowsTwoPages` property to `false` in order to show only one page at a time. Finally, we set the `pages` property with an array containing both the views, as shown in the following code:

```
var pageflip = PageFlip.createView({
    transition: PageFlip.TRANSITION_CURL,
    transitionDuration: 0.3,
    landscapeShowsTwoPages: false,
    pages: [ orangePage, bluePage ]
});
```

There are other transition effects which we can choose from, such as those shown in the following table:

Transition value	Description
`Ti.PageFlip.TRANSITION_FLIP`	When transitioning, the pages will flip from their spine
`Ti.PageFlip.TRANSITION_SLIDE`	When transitioning, the pages will slide on and off the screen
`Ti.PageFlip.TRANSITION_FADE`	When transitioning, the pages will fade between each other
`Ti.PageFlip.TRANSITION_CURL`	When transitioning, the pages will curl between each other from their spine

We add the `pageflip` component containing our two views to the window, and then open the main window just as we normally do, as shown in the following code:

```
win.add(pageflip);
win.open();
```

Making sure the pages actually turn

Let's do our very first test using a native module, by clicking on the **Run** button from the **App Explorer** tab.

The following screenshot shows the output screen on our first run, when swiping our finger across the first (orange) view:

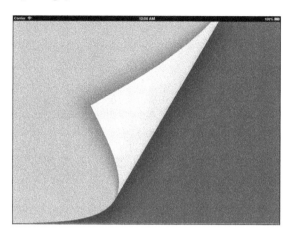

Even at this preliminary stage, we can already see that it is working. And it doesn't take much imagination to see that the addition of extra pages can be done easily since all the basics are in place.

Orientation

Since we will be using different UI components in the application, we need to ensure that it will look the same for everyone. Therefore, we decided to choose one orientation and stick with it (for this application at least). So, we need to prevent the device's orientation from changing when a user rotates his/her device.

This is relatively simple; we can do this by modifying the `tiapp.xml` file and by leaving only `Ti.UI.LANDSCAPE_LEFT` and `Ti.UI.LANDSCAPE_RIGHT` in the `orientations` list. This will force the device's orientation to support only landscape mode, as shown in the following code:

```
<iphone>
    <orientations device="ipad">
        <orientation>Ti.UI.LANDSCAPE_LEFT</orientation>
        <orientation>Ti.UI.LANDSCAPE_RIGHT</orientation>
    </orientations>
</iphone>
```

We could also have achieved the same result by setting those same constants to the `orientationModes` property while creating the window. It would look similar to the following code snippet in the `app.js` file:

```
var win = Titanium.UI.createWindow({
    title: 'Interactive eBook for iPad',
    orientationModes: [ Ti.UI.LANDSCAPE_LEFT,
                        Ti.UI.LANDSCAPE_RIGHT ]
});
```

While this method also works, it is not always recommended when developing an application with many windows. Otherwise, you will have to define these properties on every single window. This property is used more in the context of the exception than the rule. For example, a photo gallery should be displayed in landscape mode, while the rest of the application should be displayed in portrait mode.

Why not just LANDSCAPE?

We have seen the use of `LANDSCAPE_LEFT` and `LANDSCAPE_RIGHT` in the preceding section. But why not simply specify `LANDSCAPE` and be done with it? And what do these `LEFT` and `RIGHT` attributes actually mean?

The answer to these questions is quite simple actually. There is no attribute such as `LANDSCAPE`, because on iPad, there is no single landscape mode to rule all the components. When a user is holding his or her iPad sideways, he/she will either have the **Home** button to the left (`LANDSCAPE_LEFT`), or to the right (`LANDSCAPE_RIGHT`). This allows us to be more flexible in cases where we absolutely need the user to have access to the **Home** button in a certain way.

For our e-book application, there is no point in locking the orientation of the device, as long as it is running in the landscape mode. Therefore, we used both constants so that the users can rotate their iPads while still using our application.

The rich HTML page

Now that we have managed to test our application using the two empty views, let's replace those with the real content. We want our first page to provide a rich experience in terms of fonts and layout. For this case, HTML is a lot more suitable (and powerful) to achieve our goal.

But before we dive into it, we must think in terms of the code structure. If we create contents of every single page in the `app.js` file, it may prove to be cumbersome down the line. It will make the code harder to read and less maintainable. Therefore, we will create a new file called `webpage.js` that will contain all the code related to this e-books page.

The code is pretty straightforward since it contains a single function that creates a WebView using `Ti.UI.createWebView`, and assigns its `url` value with a local HTML file. We could have used a remote URL, but since we want to have an autonomous application, we use local HTML files.

```
function WebContentPage() {
  var htmlPage = Ti.UI.createWebView({
    url: 'webpage.html'
  });
```

The following function simply returns the newly created view:

```
  return htmlPage;
}
```

The reason for using the function instead of simply declaring the `htmlPage` variable is because we want to use the CommonJS pattern to instantiate our objects. In practice, this means that the file contents won't be loaded until they are actually instantiated (using the `require` function).

In order to make this new file callable, we export the function's name using the following statement. We need not worry about the inner workings as of now. But just be aware that another component can only use what is actually exported. All that is not exported is considered as private.

```
module.exports = WebContentPage;
```

We now have a rich `WebView` that brings in all of the functionalities of a browser.

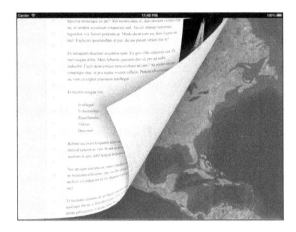

The map view

For our second page, we will be using a map view to display a satellite image of a designated area. Just as for the first page, we will create a new file called `mappage.js`. This new file will contain a function that will be used to instantiate the map view.

We can create a map view using the `Ti.Map.createView` function, and save its reference in the `map` variable. We want it to show a satellite view and show some nice animation when centering to the location. We need the map region to adapt to the application's aspect ratio; we don't want the user to be able to move around the map as it would conflict with the page turning. Lastly, we set the `region` property with the property GPS `latitude` and `longitude`.

Notice the `latitudeDelta` and `longitudeDelta` attributes. They stand for the amount of distance displayed on the map, measured in decimal degrees.

```
function MapContentPage() {
  var map = Ti.Map.createView({
    mapType: Ti.Map.SATELLITE_TYPE,
    animate: true,
    regionFit: true,
    userLocation: false,
    touchEnabled: false,
    region: {
```

```
        latitude: 48.8587011132514,
        longitude: 2.2942328453063965,
        latitudeDelta: 0.01,
        longitudeDelta: 0.01
    }
  });
```

We return the `map` variable in order to use it later:

```
    return map;
}
```

For the function to be accessible in a `CommonJS` context, we must export it using the following code:

```
module.exports = MapContentPage;
```

We now have a working satellite map view, which will be a nice addition to the book, as shown in the following screenshot:

The video player

Our third and last page will be showing a full-fledged video player. Titanium provides us with a `Video Player` component. We use a local file contained in the `Resources` directory as the URL. Just as for the `WebView` component, we could have used a remote file. But keep in mind that performance may become an issue during playback, depending on the quality of the connection.

We want the video player to have a dark background and give the user access to the video controls (**Play/Pause** button, **Volume** control, and **Full Screen** toggle). We also want the video to occupy the maximum space on the screen, while maintaining its aspect ratio (we don't want the image to stretch).

Finally, we don't want the video to start automatically as soon as it is loaded. The user will have to manually start the video playback.

```
function VideoContentPage() {
 var embeddedVideo = Titanium.Media.createVideoPlayer({
    url: 'videopage.mp4',
    backgroundColor: '#111',
    mediaControlStyle: Titanium.Media.VIDEO_CONTROL_DEFAULT,
    scalingMode: Titanium.Media.VIDEO_SCALING_ASPECT_FIT,
    autoplay: false
  });
  return embeddedVideo;
}
```

As always, we export the function in order to use it later:

```
module.exports = VideoContentPage;
```

With this, we can now view the video files directly by simply flipping to the page.

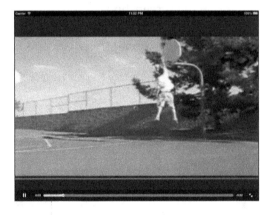

Final assembly

Now that all of our e-book pages have been created, it is now time to assemble them all for the final version. To do this, we load each file as `CommonJS` modules using the `require` function. Notice that the parentheses right after will give the effect of calling the function as soon as it is loaded.

Our `app.js` file should now look something similar to the following code snippet:

```
var page1 = require('webpage')();
var page2 = require('videopage')();
var page3 = require('mappage')();
var pageflip = PageFlip.createView({
   ...
  pages: [ page1, page3, page2 ]
});
```

Summary

In this chapter, we learned what a Titanium Native module actually is. We downloaded one from the marketplace, installed it, and configured it into our project. We learned how to instantiate a native module through code and interact with it as if it was a regular object. We learned how to specify which orientation will be supported by our application, and lock the ones we don't want to use accordingly.

We also learned how to use richer View components, such as `WebView` and `VideoPlayer`.

Last but not the least, we learned how to use a map view, center it to a certain set of GPC coordinates, and display the nearby region.

5
You've Got to Know When to Hold 'em

This chapter will take us through the development of a Stock Portfolio mobile application. It will allow users to organize their stocks as well as their respective quantities. From there, users will be able to define an amount of money they want to earn through their investments. The application will then allow on-demand retrieval of the latest stock prices from the Internet and then calculate the total portfolio's value. The application will then display how far (or close) the user is from his/her objective. When the portfolio's total value reaches the objective, the application will recommend selling the stocks.

This new application is different in terms of navigation, since it has two windows. The main window will show the investment progression using a progress bar component as well as labels and buttons for interaction. The second window will show all the stocks from the portfolio, as well as text fields to allow the users to change some settings. The stock prices will be retrieved using HTTP calls. Settings, and the portfolio itself will be persisted on the device using Titanium's properties functionality.

In this chapter, we will cover the following concepts:

- Navigating back and forth between different windows
- Using the progress bar component
- Retrieving online information using HTTP
- Creating different layers for different parts of the application (model, services, and UI)
- Setting a default UI unit for the entire project
- Creating forms using text fields and labels
- Creating custom table view rows
- Using global events

Creating our project

Since we have a lot on our plate, let's dig in right away and create our new project. To do this, navigate to the **File | New | Mobile Project** menu from Titanium Studio and fill out the wizard forms with the following information:

Field	Value to be entered
Project Template	Default Project
Project name	`KennyStock`
Location	You can either: • Create the project in your current workspace directory by checking the **Use default location** checkbox • Create the project at a location of your choice
App Id	`com.packtpub.hotshot.kennystock`
Company/Personal URL	`http://www.packtpub.com`
Titanium SDK Version	By default, the wizard will select the latest version of Titanium SDK. This is recommended. (as of this writing, we are using Version **3.1.3.GA**).
Deployment Targets	Check the **iPhone** and **Android** options.
Cloud Settings	Uncheck the **Cloud-enable this application** checkbox.

Project creation is covered with in extensive detail in *Chapter 1, Stopwatch (with Lap Counter)*, so feel free to refer to this section if you want more information regarding project creation.

The application structure

JavaScript doesn't impose much structure when it comes to file location and how the files are grouped. Titanium is no exception to this rule as the only things it imposes are that the files must be contained in the `Resources` directory, and that `app.js` is the first file that is executed.

But that doesn't mean we can't structure our files as a good practice. There are various approaches we can follow, and all share some good qualities. For this project (and some other ones later in this book), we will use this very simple directory structure for all our code under the Resource directory:

Directory	Contents
/model	Models that will act as a representation of real-life objects that will be used in our application.
/service	Code that provides access to different services (database, device storage, communication, geolocation, and so on).
/ui	All of our user interface components, whether they are reusable controls or entire windows.

As this is a fairly more complex application, it makes little sense in trying to fit the entire basecode in the app.js file. Therefore, each window will have its own individual file. This will keep the whole thing contained and easier to read.

One additional advantage of separating our source code is that it will allow us to load each file on demand, meaning that all the resources contained in a file (a window and all its children components in this case) will not be loaded into memory until it is needed. This is done using the require function, which is part of a specification known as CommonJS. Specifically, that part of the specification is called **Modules**. There are many implementations of CommonJS out there for different platforms (including Titanium). As this is not in the scope of this book, just see this as a way of loading a JavaScript code from another file, and then access objects and functions from this file as you would from any other object.

 While the require and Ti.include functions seem alike, they are really quite different. The require function will load the file at runtime only when the interpreter reaches the line where it is called, while the Ti.include function will essentially copy the entire contents of this file and merge it with the current code before any execution has taken place. This is quite different as the entire code is loaded irrespective of whether the developer chooses to use it or not.

The Ti.include function was introduced in the early versions of Titanium SDK, but now that Titanium supports CommonJS, it is recommended that we use the require function.

The main application window

Our main application window will be comprised of a progress bar that will indicate how far we are from our objective. This same progress bar will have a label on each end to indicate the minimum and maximum value. There will also be a much larger label right at the center of the window which will serve as the recommendation (whether to hold on to our stock or sell it).

Finally, there will be buttons on each window's bottom-corners, one to access the portfolio management window and a second one to refresh the stocks by retrieving their latest selling price online:

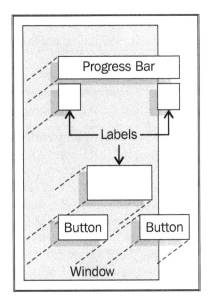

Moving on to the code

As mentioned earlier, our main application window will be contained in its very own JavaScript file. So, in Titanium Studio, we will create a new file by navigating to the **File** | **New** | **File** menu. We will name it ApplicationWindow.js and it will be located in the Resources/ui directory.

In order to be able to load the application window, we need to define a function that will return our window. Let's keep it empty for now:

```
function ApplicationWindow() {
}
```

Once our function is created, we must expose this function to the outside world by assigning the `module.exports` variable to the function we just created:

```
module.exports = ApplicationWindow;
```

That is all we need to make our window dynamically loadable using the `require` function. This assignment is also known as **exporting**.

Coding the user interface

We now need to create our window object that will be returned by the function we just created. So inside the `ApplicationWindow` function, we will create a new window variable called `self`. We will give it a title and a linear gradient as its background. It will go from a dark green color, starting at the top-left corner, to a much lighter green, ending at the right-bottom corner of the screen:

```
var self = Ti.UI.createWindow({
  title: 'Kenny Stock',
  backgroundGradient: {
    type: 'linear',
    startPoint: { x: '0%', y: '0%' },
    endPoint: { x: '100%', y: '100%' },
    colors: [ { color: '#002400'}, { color: '#b4ddb4' } ]
  }
});
```

We will also create a progress bar component that will show how far we are from reaching our objective. Since we don't have a specific monetary objective yet, we will settle for one dollar. Therefore, the absolute maximum value will be one (we will make this maximum value dynamic later down the road):

```
var progress = Ti.UI.createProgressBar({
  top: 55,
  width: '95%',
  height: '10%',
  max: 1
});
```

We will then add our progress bar to the window we created earlier:

```
self.add(progress);
```

At each end of our progress bar, we will have labels that will display the lowest and highest value the progress bar can attend. The leftmost label will be set at $0 and it will never change. We can then add it by declaring it in the add function, thus eliminating the need for an unused variable:

```
self.add(Ti.UI.createLabel({
    left: 5,
    top: 110,
    width: Ti.UI.SIZE,
    height: Ti.UI.SIZE,
    text: '0$',
    font: {
        fontSize: '16sp',
    }
}));
```

The rightmost label's value is subject to change, depending on what the users will enter. Therefore, we will store its reference in a variable called lblObjective for later use:

```
var lblObjective = Ti.UI.createLabel({
    right: 5,
    top: 110,
    width: Ti.UI.SIZE,
    height: Ti.UI.SIZE,
    text: '1$',
    font: {
        fontSize: '16sp',
    }
});
```

We will then add the label to our window:

```
self.add(lblObjective);
```

Our third label will be much bigger and centered on the screen. Since its value is destined to change over time, we store its reference in the `lblWhatToDo` variable:

```
var lblWhatToDo = Ti.UI.createLabel({
  text: 'HOLD',
  left: 5,
  top: 200,
  width: '100%',
  height: Ti.UI.SIZE,
  textAlign: Ti.UI.TEXT_ALIGNMENT_CENTER,
  font: {
    fontSize: '65sp',
    fontWeight: 'bold'
  },
  color: '#ffffff'
})
```

We will then need to add this label to our window:

```
self.add(lblWhatToDo);
```

We will need to create our two buttons, each sitting at the bottom-corner of the screen. These buttons won't have any visual styles per se; instead, we will rely on the background images:

```
var btnPortfolio = Ti.UI.createButton({
  backgroundImage: '/images/edit_folio.png',
  height: 26,
  width: 26,
  bottom: 8,
  left: 8
});

var btnRefresh = Ti.UI.createButton({
  backgroundImage: '/images/refresh.png',
  height: 26,
  width: 26,
  bottom: 8,
  right: 8
});
```

As with every other component previously created, we will need to add these buttons to our window:

```
self.add(btnPortfolio);
self.add(btnRefresh);
```

Now that our window is ready to be used, we must return it from our function. That way, when we call the `ApplicationWindow` function later on, it will return a new instance to the window we just created:

```
return self;
```

Sparing our fingers

Since the beginning of this book, all our user interface elements have been positioned and sized using **density-independent pixels** (we sometimes call these **points** or **dp**). We did that by suffixing `dp` to every single dimension attribute (for example, `width:'35dp'`). You may have noticed that we have not done that in this chapter.

While this is necessary when supporting multiple devices with different layouts, it can become cumbersome having to add it everywhere in our code. Also, there may be instances where we simply forget to add it. While this won't raise any errors, there will be visual discrepancies that may be hard to track down as the basecode gets larger.

A better way of doing this would be to set a default unit for the entire project. This can be achieved by adding the following property to our project's `tiapp.xml` file:

```
<?xml version="1.0" encoding="UTF-8"?>
<ti:appxmlns:ti="http://ti.appcelerator.org">

<property name="ti.ui.defaultunit">dp</property>
```

You can choose whichever unit that suits your needs:

- px for pixels
- mm for millimeters
- cm for centimeters
- in for inches
- dp (or dip) for density-independent pixels
- % for percentage

Of course, you still have the luxury of overriding this value by specifying it in the code. Your default unit can be in dp, but you can still define an element that has width: '30%'.

Preparing for our first run

Now that we have our first window set up, we need to invoke it. Similar to every Titanium application, our starting point is the app.js file. As usual, there will be default content already in this file; we will delete it all and replace it with our own.

Our bootstrap code will be comprised of a self-invoking anonymous function, meaning that it will be run automatically as soon as it is created. It also has no name; hence, it is called an anonymous function. The function will load the contents of the ApplicationWindow file into the memory using the require function, and store its reference in the Window variable. Finally, the last statement simply creates a new Window object and calls its open function in order to display it:

```
(function() {
  var Window = require('ui/ApplicationWindow');

  new Window().open();
})();
```

All filenames passed to the require function are relative to the root of the Resource directory. This is true irrespective of where you are located in the code. Also, you don't need to specify the file extension.

Let's take our first iteration for a spin by clicking on the **Run** button from the **App Explorer** tab.

Here is what we see on our first run:

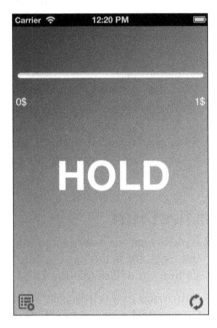

While it doesn't do much yet, it gives us a solid base to implement the rest of our navigation.

The portfolio management window

Our second window will be used to manage the contents of our portfolio as well as our application's settings such as our monetary objective. It will be comprised of two main sections.

At the top, we will find a text field where the users will enter their monetary objective. There will also be two more text fields (**symbol** and **quantity**) and a button used to add a new stock to the user's portfolio. All these text fields will be accompanied by labels next to them for clarity.

The lower section of the screen will be used to display the portfolio's stocks using a table view control. Also, at the very bottom of the screen, there will be a button that will allow to save all the settings, refresh the portfolio's data, as well as close the current window:

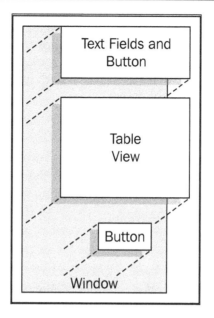

Coding what we have designed

From the top menu of Titanium Studio, navigate to **File** | **New** | **File**. This will bring up the new file-creation dialog window. We will name the file as `PortfolioWindow.js` and it will be located in the `Resources/ui` directory, just like the other window.

All our code will be contained in a function called `PortfolioWindow`, and this same function will then be assigned to the `module.exports` variable:

```
function PortfolioWindow() {
}

module.exports = PortfolioWindow;
```

The form at the top

Since all our UI controls will be contained in a window, we will create a new `Window` object and assign its reference to the `self` variable for later use. We also want this window to have a linear background going from dark brown (top-left corner) to a lighter tone of brown (bottom-right corner):

```
var self = Ti.UI.createWindow({
  title: 'Portfolio',
  backgroundGradient: {
    type: 'linear',
    startPoint: { x: '0%', y: '0%' },
    endPoint: { x: '100%', y: '100%' },
    colors: [ { color: '#752201'}, { color: '#bf6e4e' } ]
  }
});
```

Although the variable referencing our window is also named `self` (similar to the one from the application window), they are two completely separate variables with the same name. Each of them exist in their own scope; each CommonJS module loaded using the `require` function has its own scope. This provides better isolation between modules.

We will then create a label to identify the objective's text field. Once again, since we won't reference this label, we will declare it directly in the `add` function of its container:

```
self.add(Ti.UI.createLabel({
  left: 10,
  top: 30,
  width: Ti.UI.SIZE,
  height: Ti.UI.SIZE,
  text: 'Your money objective ($):',
  font: {
    fontSize: '14sp',
    fontWeight: 'bold'
  }
}));
```

We will need a text field in order to collect the user's monetary objective. We want it to have a nice rounded border and a hint text to be displayed when the field is empty, so that the user knows what is expected in this field. Since the monetary objective must be a numeric value, we need to display the number pad instead of the regular keyboard. This will better suit the user's needs and avoid invalid entry further down the line. Finally, we want the keyboard to slide down when the user hits the **Done** button:

```
var txtObjective = Ti.UI.createTextField({
  right: 15,
  top: 15,
  width: 100,
  height: Ti.UI.SIZE,
  backgroundColor: '#ffffff',
  borderStyle: Titanium.UI.INPUT_BORDERSTYLE_ROUNDED,
  hintText: 'Amount',
  keyboardType: Ti.UI.KEYBOARD_NUMBER_PAD,
  returnKeyType: Titanium.UI.RETURNKEY_DONE
});
```

We will then add this new text field to our window:

```
self.add(txtObjective);
```

Again, we will create a new label that will go next to the text field used to collect the stock's symbol:

```
self.add(Ti.UI.createLabel({
  left: 5,
  top: 100,
  width: Ti.UI.SIZE,
  height: Ti.UI.SIZE,
  text: 'Symbol:',
  font: {
    fontSize: '14sp',
    fontWeight: 'bold'
  }
}));
```

We will then create a new text field component that will be used to add a new stock to the portfolio (the short symbol we usually see scrolling on the news tickers):

```
var txtSymbol = Ti.UI.createTextField({
  left: 63,
  top: 85,
  width: 70,
  height: Ti.UI.SIZE,
  backgroundColor: '#ffffff',
  borderStyle :Titanium.UI.INPUT_BORDERSTYLE_ROUNDED,
  returnKeyType: Titanium.UI.RETURNKEY_DONE
});
```

We will then add this same field to our window:

```
self.add(txtSymbol);
```

We will require one last label that will go next to the quantity text field:

```
self.add(Ti.UI.createLabel({
  left: 140,
  top: 100,
  width: Ti.UI.SIZE,
  height: Ti.UI.SIZE,
  text: 'Quantity:',
  font: {
    fontSize: '14sp',
    fontWeight: 'bold'
  }
}));
```

Our last text field will be used to collect the quantity of this stock owned by the user. Just as with the text field for a monetary objective, we want to limit the input to just numbers using the number pad:

```
var txtQuantity = Ti.UI.createTextField({
  left: 205,
  top: 85,
  width: 75,
  height: Ti.UI.SIZE,
  keyboardType: Ti.UI.KEYBOARD_NUMBER_PAD,
  backgroundColor: '#ffffff',
  borderStyle :Titanium.UI.INPUT_BORDERSTYLE_ROUNDED,
  returnKeyType: Titanium.UI.RETURNKEY_DONE
});
```

We will then need to add this newly created field to our window:

```
self.add(txtQuantity);
```

In order to actually add a new stock to the portfolio, we will need a button. As we will do something when the user clicks on it, we keep its reference in the btnAddStock variable:

```
var btnAddStock = Ti.UI.createButton({
    right: 2,
    top: 85,
    title: 'Add'
});
```

This, in turn, will be added to our window:

```
self.add(btnAddStock);
```

The stock list

The portfolio's content will be displayed in the form of a list using a table view control. We will place it right under the form we created previously. It will span the screen's entire width, and it will occupy half the height of the screen. We won't put any content in it right now, as this will be done later down the road:

```
var stockList = Ti.UI.createTableView({
    left: 0,
    top: 170,
    width: Ti.UI.FILL,
    height: '50%'
});
```

We will also need to add the table to our window:

```
self.add(stockList);
```

Last but not the least, at the bottom of the screen, we will create a large button, which will span 80 percent of the screen's width:

```
var btnSave = Ti.UI.createButton({
    bottom: 10,
    width: '80%',
    title: 'Save Settings'
});
```

Finally, we add our last control to the window:

```
self.add(btnSave);
```

The portfolio management window is now complete in terms of UI elements. We must then return the `self` variable from our `PortfolioWindow` function:

```
return self;
```

Navigating between the two windows

While our new portfolio management window is created, there is still no way to test it, as all our application is doing right now is showing the first window. What we need to do, is put a mechanism in place for opening this new window from our main application window. For this, we will edit the `ApplicationWindow.js` file and add the following event handler to the button named `btnPortfolio`.

In our `click` event, we will load the contents of the `PortfolioWindow.js` file using the `require` function and store this content in a variable just as we did the last time. Then, we will invoke a new instance of our portfolio window, and display it using the `open` function. Finally, we want the window to have a flip transition from the left when it appears.

 The window transitions are iOS specific. Although the code won't have any impact on Android at runtime, you will see an error in the logs.

The flip transition can be implemented as follows:

```
btnPortfolio.addEventListener('click', function(e) {
    var PortfolioWindow = require('ui/PortfolioWindow');

    new PortfolioWindow().open({
        transition: Ti.UI.iPhone.AnimationStyle.FLIP_FROM_LEFT
    });
});
```

 You can put the code of the event handler just about anywhere, as long as it comes after the variable declaration. But it is usually considered a good practice to keep the event handler declarations close to where the control is declared for greater readability.

Let's see if it works

Now that the code is in place, let's run our application one more time and see if we can navigate from one window to another. Once the application is launched, click on the **portfolio** button at the bottom-left corner of the screen.

You should see our portfolio window as follows:

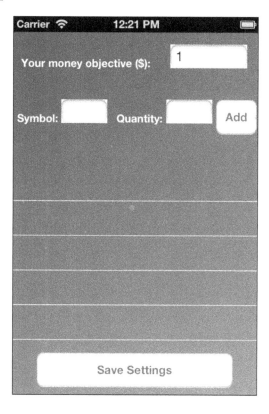

Of course, we still have a lot to do here. The controls on this window have no interaction, and most of all, once you are in the portfolio window, there is no way to get out of there, except by quitting the application altogether. Let's fix that by adding stock management and navigation to our application.

Moving stocks around

In order to have a clear understanding of how we will manage our stocks, we must first define what exactly constitutes a stock in our application. For our needs, a stock must have the following attributes:

- Its symbol, in order to distinguish it from the others

- Its price on the stock market

- How much of this stock do we own? (quantity)

Now, we need to represent the model in the code that is easy to handle. For this, we will create a new file called `Stock.js` and it will be located in the `Resources/model` directory.

Inside this new file, we will create a new function called `Stock` with two parameters. Inside the function, we assign the symbol and quantity attributes with the values passed as parameters. We will set the price to its default value of zero dollars and zero cents. Notice that those two parameter values are the same information that will be required in the forms when adding a new stock to the portfolio:

```
function Stock(symbol, quantity) {
   this.symbol = symbol;
   this.quantity = quantity;
   this.price = 0.00;
}
```

As we are used to by now, using CommonJS, we assign our newly created function to the `module.exports` variable in order to be able to access it:

```
module.exports = Stock;
```

Saving our portfolio

There are many ways we could use to save our portfolio information on the device. For this chapter, we will be using the Titanium Properties API that is specifically made for handling application-related data in property/value pairs that persist beyond application sessions. Even if the device is turned off, the data will still be available.

Instead of spreading the properties interaction code all over our application, we will wrap it in a service that we will call the preference service. We will create a new file specifically for it, named `PreferenceService.js`, and it will be located in the `Resources/service` directory.

Since we don't know if we want to expose everything from our service to the outside world (as opposed to a window), we will create a `PreferenceService` object and assign it with an empty function:

```
var PreferenceService = function() {}
```

Keeping up on our objective

We want our service to expose functions to retrieve and save our user's monetary objective. Therefore, we add a new function to the `PreferenceService` object using the `prototype` JavaScript property available to all objects, and then assign a new function that will contain our code.

Now that our function is created, we want to be able to save our objective to the device. The `Ti.App.Properties` object provides us with different functions to save many types of objects. The basic principle is quite simple; when you save a value, you do so by providing a key under which it will be stored. Later, when you need to retrieve this information, you simply have to provide this same key and the value will be returned by the getter function. If a value matching the key is not found, a default value can be specified if needed.

In this particular case, we want to save a number so we will use the `setInt` function and store its `value` under the `'objective'` key:

```
PreferenceService.prototype.saveObjective = function(value) {
   Ti.App.Properties.setInt('objective', value);
}
```

We will also expose a function that allows us to retrieve the same objective we saved previously, using the `getInt` function, by passing the same key we used while saving it. In case there is no objective saved yet (on the application's very first run), we will return a default value of 1:

```
PreferenceService.prototype.getObjective = function() {
   return Ti.App.Properties.getInt('objective', 1);
}
```

The default value could be any number we choose. We just want to keep it as simple as we possibly can.

Portfolio management

The properties also allow us to handle more complex objects such as lists. Although our application doesn't give us the ability to create stocks (yet), having functions to manage them will make things a lot easier later down the road.

Saving

As we want to be able to use the same function while we are saving the whole list and when we update one stock from the list, we will create a new function and store its reference in the `saveStocks` variable for later use. This function will save the list that was passed as a parameter, using the `setList` function under the `'stocks'` key.

Since saving or updating the stock list has a direct impact on the total value, we recalculate the entire portfolio's value using the `updatePortfolioValue` function. This function is not defined yet; we will do so very soon in this section, so don't you worry about that:

```
var saveStocks = function(list) {
  Ti.App.Properties.setList('stocks', list);
  updatePortfolioValue(list);
};
```

We now want to expose this `saveStocks` function to the service so that it is accessible from outside this file:

```
PreferenceService.prototype.saveStocks = saveStocks;
```

Retrieving

Since it makes sense to retrieve everything, we save, we will create a `getStocks` function. Just as its `saveStocks` counterpart, we will need to call it in other functions. This means that we will store its reference in a variable named `getStocks` for later use.

We will then retrieve the stock list using the `getList` function. Again, if there are no stock lists already saved on the device, the function will return an empty array as the default value:

```
var getStocks = function() {
  return Ti.App.Properties.getList('stocks', []);
};
```

We will expose this new `getStocks` function to the outside world:

```
PreferenceService.prototype.getStocks = getStocks;
```

Updating a particular stock

There will be cases where we will want to update a single stock from our list, for example, when the prices change. For this particular case, we will create an updateStock function that will pass the updated stock as a parameter.

The algorithm is pretty straightforward. First, we load the whole stock list. Then, we loop through all of them in order to determine which one to update. We do this by comparing the stock's symbols (the only way to identify them). Once we find it, we simply replace the stock from the list with our updated stock. Finally, we save our updated stock list.

```
PreferenceService.prototype.updateStock = function(updatedStock) {
  var allStocks = getStocks();

  for (var i=0; i < allStocks.length; i++) {
    if (allStocks[i].symbol === updatedStock.symbol) {
      allStocks[i] = updatedStock;
    }
  }
  saveStocks(allStocks);
}
```

How much is our portfolio worth?

One of the core features of our new application is to give the user the ability to know exactly how much money he/she has. Therefore, the service needs such a function. We will create a function named updatePortfolioValue with a list of stocks as a parameter.

To determine the portfolio's value, all we have to do is iterate over the stock list, multiply each stock's price by the quantity owned, and add this to our grand total. In order to avoid repeating the process every single time the application starts, we will save it using the setInt function under the 'portfolioValue' key for later use:

```
function updatePortfolioValue(stocks) {
  var totalValue = 0.00;

  for (var i=0; i <stocks.length; i++) {
    var s = stocks[i];

    totalValue += (s.price * s.quantity);
  }

  Ti.App.Properties.setInt('portfolioValue', totalValue);
}
```

Since all the content of this file are not contained in a single function, we need to create a new instance of our `PreferenceService` class, and it is this same instance that we will then export from our module:

```
var pref = new PreferenceService();
module.exports = pref;
```

Wiring the preference service to the UI

Now that we have our preference service in place, we can now add a functionality to our portfolio window, knowing that the changes done by our user won't be lost.

Adding a new stock to the list

Probably the most important feature in our application is the ability to add new stocks. Let's start with this one. We will add a `click` event handler on the `btnAddStock` button. Inside this handler, we will create a new `Stock` object using the symbol (in upper case), and quantity entered by the user. Once our stock object is created, we will add it to the table view underneath using the `addCustomRow` function. We will create this function in just a moment, but for now, remember that it takes two parameters, the table view in which to add the new row and the stock object containing the information we will need to display. Finally, we reset the `txtSymbol` and `txtQuantity` text field values so that the user can add a new stock if desired:

```
btnAddStock.addEventListener('click', function() {
  if (txtSymbol.text != '' &&txtQuantity.text != '') {
    var stock = new Stock(txtSymbol.value.toUpperCase(),
      txtQuantity.value);

    addCustomRow(stockList, stock);

    txtSymbol.value = '';
    txtQuantity.value = '';
  }
});
```

Creating a custom look for our table rows

In the previous chapters, we have dynamically created table view rows. But all these rows abide by the default layout provided by Titanium. While you can do many things with all the properties that the `TableViewRow` object provides, there will be times where you will really need a more custom layout. One of the most common example is the Facebook application; it has a very distinct look that is pretty far from the standard list we usually see in other applications.

We can achieve this by creating a `TableViewRow` object, and then add child controls to it, similar to what we would do with a regular view:

```
function addCustomRow(table, stock) {
    var row = Ti.UI.createTableViewRow();
```

We will then add a label to our row to display the stock's symbol. It will be positioned at the top-left corner of our row and it will have a bigger, bolder font:

```
row.add(Ti.UI.createLabel({
    text: stock.symbol,
    left: 5,
    top: 1,
    font: {
        fontSize: '15sp',
        fontWeight: 'bold'
    }
}));
```

We will also add a label to display the stock's latest price. It will be located at the top-left corner of our row and it will have a more regular font:

```
row.add(Ti.UI.createLabel({
    text: 'Latest Price: ' + stock.price + '$',
    right: 5,
    top: 1,
    font: {
        fontSize: '9sp',
    }
}));
```

We will create one last label to display how much of a particular stock is in our portfolio. The font for this label will be much smaller and it will be located right under the one we just created earlier:

```
row.add(Ti.UI.createLabel({
   text: + stock.quantity,
   right: 5,
   top: 20,
   font: {
      fontSize: '8sp',
      fontStyle: 'italic'
   }
}));
```

We will also need to keep a reference to our original stock object. This will be useful down the road when we loop or interact with the table view rows and want to determine which actual stock is related to the information displayed in the row:

```
row.stock = stock;
```

Once our new custom row is complete, we append it to the table view just as we would do with any other row:

```
table.appendRow(row);
}
```

One more thing before we can test

We now have all our code set up to add and display new stocks in our portfolio window; we can now take our application for a third run. But before we do that, we need to load the modules we will be using in this window to the memory using the `require` function.

We need the `Stock` module that will be used to handle our model objects. We will store its content in the `Stock` variable. We also need the `PreferenceService` module in order to save and retrieve our objective and stock list. We will store its content in a variable named ps, as this will make the code easier to read:

```
var Stock = require('model/Stock');
var ps = require('service/PreferenceService');
```

We can now run our application, load the portfolio management window, and add stocks to our list. We should see something similar to this:

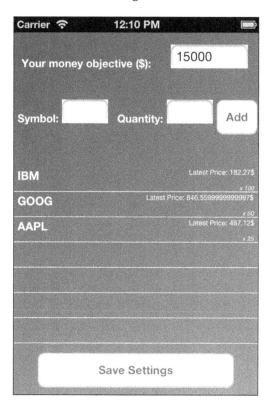

This test run demonstrates that we can now create `Stock` objects based on our model, and that we can display those same objects in a table view using a custom look and feel.

Saving all that

Now that we can set the objective and define the stocks from the portfolio, we need to add the ability to save all this information. Therefore, we will add a `click` event handler on the `btnSave` button:

```
btnSave.addEventListener('click', function() {
```

The first thing we will do is to save the objective by calling the `saveObjective` function we defined in the preference service. If the field was left blank, we will set it to `0`:

```
ps.saveObjective(parseInt(txtObjective.value) || 0);
```

We will then need to save our stock list. To achieve this, we will declare an empty array to store our list. We will then iterate through the table view's rows in order to retrieve the original stock object we attached earlier when constructing the list. Once we get a hold of the original object, we will add it to our `stocks` array. Finally, we save our portfolio using the `saveStocks` function from the preference service:

```
var stocks = [];

var section = stockList.data[0];

for (var i = 0; i <section.rowCount; i++) {
  var row = section.rows[i];

  stocks.push(row.stock);
}

ps.saveStocks(stocks);
```

> In order to loop through all the table rows, we had to get the first section reference. That's because the table view component always has at least one section, even if we don't explicitly define one.

Since the application's settings have now changed, we now need to inform the parent window that this has changed. To achieve this, we will use a global event.
A global event works just as any other event, only we can add an event listener from wherever in the code:

```
Ti.App.fireEvent('app:portfolioChanged');
```

Finally, we close the current window in order to return to the main application window:

```
self.close({
  transition: Ti.UI.iPhone.AnimationStyle.FLIP_FROM_RIGHT
});
});
```

What if there are stocks already?

For cases where the user has already defined his/her objective and portfolio, we will need to prefill the window with the saved information. To achieve this, we will edit the `PortfolioWindow.js` file and add the following code right before the last statement of the `PortfolioWindow` function:

The code will assign the objective value to the txtObjective text field. It will also call the addCustomRow function for each stock previously saved:

```
...
txtObjective.value = ps.getObjective();

var stocks = ps.getStocks();

for (var i=0; i < stocks.length; i++) {
  addCustomRow(stockList, stocks[i]);
}

return self;
```

Retrieving the stock values from the Internet

The very last piece of our puzzle will be to retrieve the stock's latest price from the Internet. For this, we will encapsulate this feature into its own service. We will create a new file named OnlineQuotesService.js and it will be located in the Resources/service directory.

Inside this file, we will load into memory the preference service using the require function, and store its content in a variable named ps for later use. We will also create an empty function named OnlineQuotesService that will represent our service:

```
var ps = require('service/PreferenceService');

function OnlineQuotesService() {}
```

Knowing where to get it

Before we do any coding, we need to determine how we will be retrieving the information. There are many online web services to choose from. But for our specific purpose, we will be using the Markit Stock Quote API. Its main advantages are the fact that it is free and that the return format used by the API is JSON, which will eliminate the need for any parsing and translation.

You can find all the documentation regarding this API at http://dev.markitondemand.com/#stockquote. You can look it up if you want to learn more about its inner workings.

Let's see what we get from it

The Stock Quotes API is pretty easy to use; all we have to do is call the designated URL and pass the stock's symbol as the only parameter. This will return a JSON formatted string containing the information we need as well as a few extra ones.

The following code is an example of the API query for a major computer company using the `curl` command-line tool:

```
$ curl http://dev.markitondemand.com/Api/Quote/json?symbol=AAPL
```

```
{"Data":{"Status":"SUCCESS","Name":"Apple Inc","Symbol":"AAPL","LastPr
ice":439.9,"Change":0.0199999999999818,"ChangePercent":0.004546694553
05368,"Timestamp":"Fri Jan 25 15:59:59 UTC-05:00 2013","MarketCap":41
3091614200,"Volume":2342602,"ChangeYTD":532.1729,"ChangePercentYTD":-
17.3388949343343,"High":456.22,"Low":437.38,"Open":451.74}}
```

While the formatting doesn't render the result much readable, we can already identify the following:

- The returned information is contained in a property named `Data`
- We can see that the latest trading price is listed under the `LastPrice` property

Retrieving one stock

Now that we are sure the API will return the required information, we will create a new function called `getLastPrice`. This function will take one stock object as an input parameter. We will then use the stock's symbol in order to build the API's URL:

```
function getLastPrice(stock) {
  var url='http://dev.markitondemand.com/Api/Quote/json?symbol='
    + stock.symbol;
```

Titanium provides us with an object specifically designed to make asynchronous HTTP requests called `HTTPClient`. For those already familiar with the `XMLHttpRequest` specification (widely known as AJAX), you will feel right at home with its behavior. For those less familiar with it, all you need to know is that there are three steps in making a typical HTTP request.

Creating an HTTPClient object

We will create a new `HttpClient` object using the `createHTTPClient` function. We will then assign it the following three properties:

- The `onload` callback function will be called when the server will return the content (without any error):

  - The response's entire content is contained in the `this.responseText` variable. Since this response string abides by the JSON format, we can easily transform it to a valid JavaScript object using the `JSON.parse` function.

  - From there, we can retrieve the `LastPrice` attribute and assign it to our stock (notice that we rounded the number to two decimals for presentation purposes).

  - Once the stock's price has been updated, we will fire a global event to inform the application that the stock has been updated. The event will also carry the updated stock object. This is necessary in our case since the calls to the server are asynchronous, and we have no way for certain to know when the server response will arrive.

- The `onerror` callback function will be called in case the server returns an error. In this case, we will log the error message to the console.

- The timeout is kept to five seconds in case there is no response after this delay, pretty much like you would do in an Internet browser:

```
var xhr = Ti.Network.createHTTPClient({
  onload: function(e) {
    var quote = JSON.parse(this.responseText).Data;

    var newPrice = quote.LastPrice;
    stock.price = Math.round(newPrice * 100) / 100;

    Ti.App.fireEvent('oqs:stockUpdated', stock);
  },
  onerror: function(e) {
    Ti.API.error(e.error);
  },
  timeout: 5000  /* in milliseconds */
});
```

Opening the HTTPClient object

Once the `HttpClient` object is created, we will open the connection by specifying which method we want to use when communicating with the URL. In this case, we will use `GET` since the URL content is enough for the server. If we ever need to send data to the server (authentication is a common example for this), we would then use `POST`.

```
xhr.open("GET", url);
```

Sending the request itself

Now that the `HttpClient` object is properly created and that the request is opened, we will need to actually send the request to the server:

```
    xhr.send();
}
```

 The send function doesn't return anything. This is because the callback functions (`onload` and `onerror`) will handle the server responses.

Retrieving all stocks in a single call

We will finally expose a function (using the `prototype` property) that will allow us to retrieve the last selling price for every stock contained in the portfolio in a single call. We will first get all the stocks using the `getStocks` function from the preference service. We will then iterate over each stock and call the `getLastPrice` function for the said stock:

```
OnlineQuotesService.prototype.fetchValues = function() {
  var stockList = ps.getStocks();

  for (var i=0; i <stockList.length; i++) {
    var s = stockList[i];
    // Update the Latest Stock Price
    getLastPrice(s);
  }
}
```

Just as we did for the preference service, we will need to create an instance of the
`OnlineQuotesService` class and then export it so that the functions declared in this
file are visible to the outside world:

```
var service = new OnlineQuotesService();
module.exports = service;
```

Calling the web service

Many of you may have noticed that while we have written the necessary code
to retrieve all the stocks' prices, we haven't used this code anywhere in the
application yet. We will address that problem by editing the `ApplicationWindow.js`
file, and by adding a `click` event handler to the `btnRefresh` button (the one in
the lower-right corner).

We will first update the objective label located at the end of the progress bar,
based on the value returned by the preference service. We will then load the
`OnlineQuotesService` module into memory using the `require` function, and store
its content in a variable named `oqs`. Finally, we will call the `fetchValues` function to
update all the stock prices:

```
btnRefresh.addEventListener('click', function(e) {
  lblObjective.text = ps.getObjective() + '$'

  var oqs = require('service/OnlineQuotesService');

  oqs.fetchValues();
});
```

As with every event listener, you can place it anywhere you choose in the
`ApplicationWindow` function as long as it is defined after the button's declaration.

The newly added event handler will raise an error at runtime, unless you load the
preference service into memory while loading the main application window. To do
this, we will add the following line at the very top of the `ApplicationWindow.js` file:

```
var ps = require('service/PreferenceService');
```

Handling when stocks are updated

Earlier, in the online quotes service, we fired a global event every time we retrieved the latest price of a stock. We must now implement some code to be executed when such an event is fired. Inside the `ApplicationWindow.js` file, right at the end of the `ApplicationWindow` function, we will add a new event listener for the `oqs:stockUpdated` global event:

```
Ti.App.addEventListener('oqs:stockUpdated', function(stock) {
```

The first thing we will do is update the stock using the preference service:

```
ps.updateStock(stock);
```

We will then update the progress bar's attributes using the newly calculated values:

```
progress.value = ps.getPortfolioValue();
progress.max = ps.getObjective();
progress.show();
```

Finally, we will determine if our user has reached his/her objective. If that is the case, we change the big label's value to SELL:

```
if (progress.value < progress.max) {
  lblWhatToDo.text = 'HOLD';
} else {
  lblWhatToDo.text = '! SELL !';
}
});
```

Making the whole thing more fluid

The application is now fully functional and is now usable as a whole. But there are two little extra things we can do to make the whole user experience more complete and polished.

Updating stocks when the portfolio has changed

When the user changes the content of the portfolio (or the value of the objective), it makes sense to refresh the stock prices. This will then inform the user immediately how close he/she is from his/her objective.

Instead of implementing a lot of code, we can leverage the existing code already in place of the `btnRefresh` button. To achieve this, we will fire the `click` event on the button itself. Inside the `ApplicationWindow.js` file, we will add the following event listener:

```
Ti.App.addEventListener('app:portfolioChanged', function() {
  btnRefresh.fireEvent('click');
});
```

Updating stocks as the application starts

One very last thing we can add to make the whole application feel more polished would be to update all the stocks' values every time the application window is opened. At the end of the `ApplicationWindow.js` file, we will add a listener for the `open` event of the window object called `self`:

```
self.addEventListener('open', function() {
  btnRefresh.fireEvent('click');
});
```

We now have a fully-functional investment mobile application that is compatible with both iOS and Android. This application is different from the other applications we have developed so far because it relies on online information to provide a more complete experience.

Summary

In this chapter, we learned how to develop a more complex application that has two standalone windows. We have learned how to implement navigation between those two windows. We also structured our code in order to separate the model, the service, and the user interface layer. On the user interface front, we learned how to use new components such as the progress bar, text fields, as well as custom table view rows to present our information.

On the more generic front, we have learned to develop an entire application that relies on information provided by a web service. We also learned how to set the default UI unit for an entire project from the `tiapp.xml` file. Finally, we also learned how to implement global application events.

In the next chapter, we will be developing a full-fledged video game.

6
JRPG – Second to Last Fantasy

This chapter will take us through the development of a basic, yet functional mobile game. The game will allow us to control a hero shown from the **top-down perspective** that will roam around a map which is bigger than the screen itself (similar to a Japanese RPG). Simply put, when the hero reaches the edge of the screen, the map will scroll on the screen in order to allow the whole game world to be explored.

By the end of this chapter, we will have a barebones game and will have covered the following concepts:

- Incorporating a game engine into our project
- Creating the game's world using a map editor
- Working with graphics, sprites, and animations
- Loading game assets into a game
- Interacting with the game using touch events

Creating our project

As with every application we have created so far, we will create our new project by selecting the **File | New | Mobile Project** menu from **Titanium Studio**, and fill out the wizard forms with the following information:

Field	Value to enter
Project Template	**Default Project**
Project name	`SecondToLastFantasy`
Location	You can either:
	• Create the project in your current workspace directory by checking the **Use default location** checkbox
	• Create the project in a location of your choice
App Id	`com.packtpub.hotshot.secondtolastfantasy`
Company/Personal URL	`http://www.packtpub.com`
Titanium SDK Version	By default, the wizard will select the latest version of Titanium SDK. This is recommended. (as of this writing, we were using Version **3.1.3.GA**)
Deployment Targets	Check **iPhone**, **iPad**, and **Android**
Cloud Settings	Uncheck the **Cloud-enable this application** checkbox

Project creation is covered in more extensive detail in *Chapter 1, Stopwatch (with Lap Counter)*, so feel free to refer to the same section if you want more information regarding project creation.

The game's basic design

Our game's entire navigation will be done in a single window. As for the user interface, we will not be relying on the native controls such as `View`, `Button`, and `Label` as we have done with our previous applications. Instead, we will be using graphical assets that will run on top of a game engine.

Everything will be shown through a Game Scene, which will determine what we see on the screen. As we mentioned earlier, our hero will be able to roam on a map that is much bigger than the game scene itself. It is the Game Scene's role to scroll the map in order to display the part that we need to display. Those who are familiar with the `MapView` component will be able to notice the similarities.

Over that map, we will have our hero that will be a graphical component on to itself and will be able to mode all over the screen. Finally, there will be a virtual gamepad that will give our users control over our hero allowing him to move around the map.

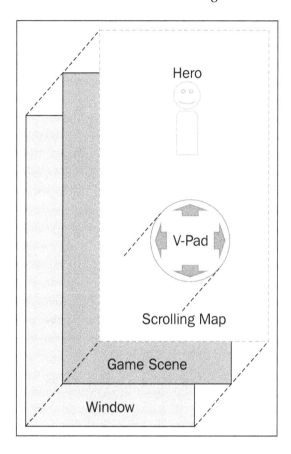

We need an engine!

While we could use standard views and image views to create our graphics and move them around the screen, this approach would be problematic in many aspects. The first one would be performance; even when it's empty, a `View` is still a pretty heavyweight native component. Another aspect would be related to animation and visual effects. While Titanium provides us with functions to stretch, rotate, move, or even animate the images, such functions are not well-suited for high performance graphics.

Therefore, we will rely on a native game engine called `quicktigame2d`. This engine will provide us with the functionality that we expect from a gaming engine.

It includes features such as:

- Game view and scene management
- Sprites and transformations
- Animations
- Maps
- Particles

We will go over most of these items throughout this chapter, but for now, all we need to keep in mind is that the game engine is basically a native module. This will give us the ability to manipulate low-level graphics from our Titanium code.

Where can we get it?

The `quicktigame2d` native module (our game engine) is available for both iOS and Android, which makes it very flexible when developing a game targeted for multiple platforms.

There is an official download page that lists all of the versions ever made from this engine since it has been in development. You can access this list at the following URL:

`http://code.google.com/p/quicktigame2d/downloads/list`

For our project, we will be using Version 1.2 of the module (whether it is for iOS or Android). In order to make sure that the right files are used throughout this chapter, please refer to the following table containing the appropriate download links:

File description	Download link
QuickTiGame2d module 1.2 for iOS	`http://bit.ly/Z8YsEw`
QuickTiGame2d module 1.2 for Android	`http://bit.ly/11rDRs7`

If you go to the game engine's website, you will notice that the team that makes the engine has been acquired by a larger company called **Lanica, Inc**. This means that the engine will no longer be supported in this incarnation. But we can still use the latest versions and it is very well-suited to our needs.

Installing our module

Now that we have retrieved our module, we need to add it to our project. Just as we did in *Chapter 4, Interactive E-book for iPad*, we will copy the module's archived file (`com.googlecode.quicktigame2d-iphone-1.2.zip` for iOS or `com.googlecode.quicktigame2d-android-1.2.zip` for Android) to our `/Resources` directory, and then build our project. From there, the Titanium build scripts will extract the module files and copy them to the appropriate directories automatically. To do this, all we need to do is to simply run the application on the desired target platform by selecting the **Run** menu.

> We can install many implementations of the same module in the same project. This brings great flexibility since we can have the same native behavior, even though the underlying implementation differs. This is similar to a standard Titanium application.

Now that our module is installed, we need to make sure that it is properly configured by adding the proper XML section in our `tiapp.xml` file. We simply add the following lines that identify the module, its platform, as well as its version.

If we were planning on developing an iOS-only game, we would need to add the following section to our `tiapp.xml` file:

```
<modules>
  <module platform="iphone" version="1.2">
    com.googlecode.quicktigame2d
  </module>
</modules>
```

On the other hand, if we were planning on developing an Android-only game, we would need to use a similar XML snippet, where only the `platform` attribute will change:

```
<modules>
  <module platform="android" version="1.2">
    com.googlecode.quicktigame2d
  </module>
</modules>
```

Since we want our game to be played on as many platforms as possible, we will declare both the modules in the same section with exactly the same module identifier (`com.googlecode.quicktigame2d`) and version number, and the appropriate implementation will be loaded depending on the target platform:

```
<modules>
  <module platform="iphone" version="1.2">
    com.googlecode.quicktigame2d
  </module>
  <module platform="android" version="1.2">
    com.googlecode.quicktigame2d
  </module>
</modules>
```

One useful way to validate the module, if it is configured properly, is by looking at the **Modules** section from the **Overview** tab in **Titanium Studio**. In the following screenshot, we can clearly see that the module was properly installed and configured for Android and iOS. In cases where the module is not properly installed, a warning icon will be shown at the beginning of the line. On the other hand, if the module is properly installed but defined with the wrong version, the corresponding table cell will be shown with a red background:

The map (before we code a single line)

Now that we have our game engine in place, we can start building our world. All this is not done by code. It does not mean that it is not feasible using only code, but the amount of work would be too much. Therefore, we will be using graphical tools and assets to achieve our goal.

The right tool for the right job

As we have all probably realized by now, **Titanium Studio** does not provide any toolset to edit video game maps (if it did, it would have been surprising). Therefore, we will be using a dedicated tool called **Tiled map editor**, or **Tiled** for short.

Tiled is a free tool that was created by *Thorbjorn Lindeijer* in 2008. Since then, it has been made available to the open source community under open source license, which means that we can use it freely for our projects.

To get it, we need to go to the map editor's official website at http://www.mapeditor.org. Then, go to the **Download** section. It is available on most of the platforms, such as Windows, Mac, and even on different versions of Linux. Once we have downloaded and installed the software, we can launch it and we will be presented with the main window that looks similar to the following screenshot:

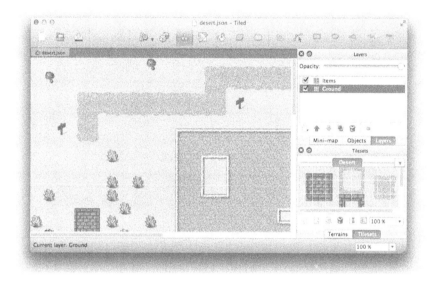

While this section is not a complete guide on how to make game maps, we will cover briefly how to make a standard map and how to export this map to a format that will be compatible with our game later.

Creating the map

A map is basically made out of tiles, (think of it as a grid) where our hero will be able to move around. Each square on this grid is a filled background image; it is the map designer's duty to make sure that all these single tiles graphically match, so that the whole map has a nice, unified look and feel, similar to a puzzle.

We will now create our game's map by clicking on the **File | New** menu from the Tiled map editor. We will keep the default orthogonal orientation and set our map to be 40 tiles wide by 40 tiles high. Finally, we want each tile to be 32 x 32 pixels, as shown in the following screenshot:

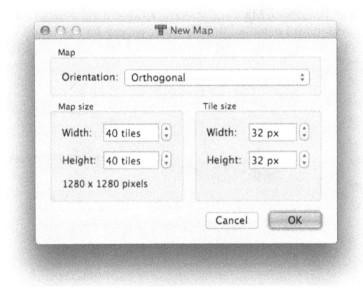

 The **New Map** creation window has a label in the **Map size** section that tells us the overall dimensions of the map. This can be useful in real life scenarios, where designers do not work with the square maps. (Sometimes the tiles are not even square shaped.)

Tilesets

Now that we have our blank slate, we need something to fill those tiles. One, less than optimal, approach would be to have a small image for each possible tile. Such an approach has several drawbacks. The first one is obvious; we would have to refer every single small image in the editor in order to place them on the map. We would also have to instantiate all of those images later on in our code, thus having a lot of repetitive code.

Another more effective approach, would be to have all our tile images in a single file, so that we can load them only once whether it is with the Map editor or in our code. Such an image is usually called a **tileset** (the term is rather explicit in terms of what it does). There are tons of free tilesets available online which you can use, but the following diagram shows the one that we will be using for our game:

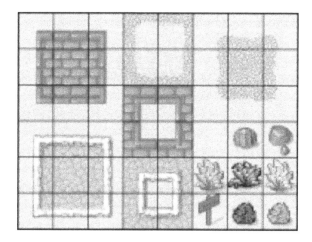

Notice the dark lines surrounding each tile; this is merely a separation margin between each tile. While being optional, it becomes easier for the designers to know where a tile ends and another one starts. This margin can be of any size since most modern tools can dynamically adapt, depending on the margin size.

 You can use any tileset you want, but we took the liberty of providing one in the application's code available on the public GitHub repository. (Refer to the *Appendix, References*).

Using the tileset

For using our tileset later in our code, we will create an `assets` directory right under the `Resources` directory. This new directory will contain all the graphical and technical assets used by our game.

We will now import our tileset into our editor by selecting the **Map | New Tileset...** menu from our map editor. We will name it as `Desert`, since we want to stay with the theme; we will then select the tileset file we just copied in our `assets` directory. Finally, we will define the size of our tiles so that the program knows how to divide the image into smaller ones. We will set our tiles to 32 pixels wide by 32 pixels high; we will also set the margin and spacing to one pixel in order to leave out the black lines around each tile, as shown in the following screenshot:

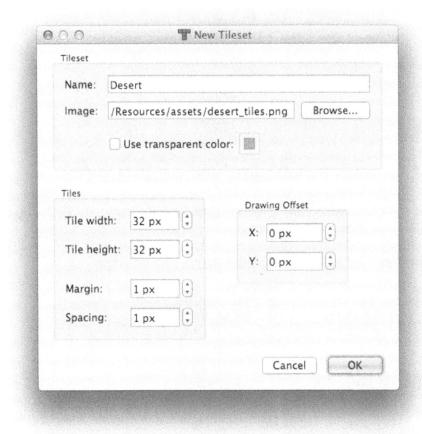

Thereafter, we will be able to select tiles from the bottom left-hand corner of the editor window and use them to build our map. Keeping in mind that our game will rely on a single tileset, it doesn't take a lot of imagination to understand that most modern games have many tilesets.

Using layers

If you take a closer look at the tileset diagram, you will notice that the tiles can be split into two categories. One represents the ground part of the map, such as sand, rocky roads, and pavements. The second one represents items that you would see on the ground, such as rocks, bushes, and sign posts.

We will now separate those into two different layers in order to display or hide them at will. To achieve this, we will add a new layer to our map by selecting the **Layer | Add Title Layer** menu. Once created, we will rename View to Items and will also rename the default layer to Ground, so that we can distinguish them more easily.

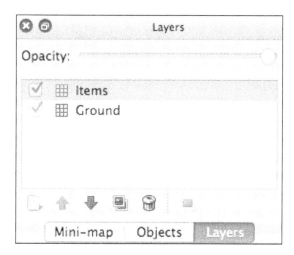

Why layers?

Layers are very useful when you want to design maps with a lot of content, but don't want everything on the same plane. If you have ever used a graphics tool such as Photoshop, you will be familiar with this concept.

One common example is to have:

- One layer for the ground
- One for the décor such as houses, bridges, and so on
- Another one for the trees, rocks, and so on

But in this chapter, we will concentrate only on two layers. Now that we have created our layers, we will be able to apply each item to its corresponding layer simply by checking and unchecking the ones we are working on in the **Layers** section of the map editor.

Speaking the same language

Now that our map is complete, we will convert it into a format understandable by Titanium as well as the `quicktigame2d` module. Luckily, Tiled provided us with an export feature that will allow us to export our map to the JSON format. There are other supported formats, but since `Titanium` reads JSON natively, this makes it the best choice for our needs.

We will now export our map by selecting the **File | Export As...** menu. From the save dialog window, we will select the `assets` directory as our destination and select the JSON format as the file type, as shown in the following screenshot:

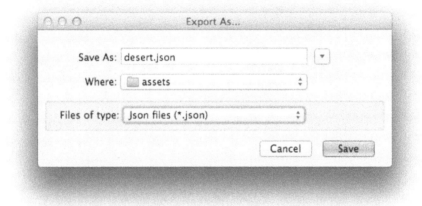

 There might be cases where the export process will add the complete tileset path in the JSON file. While this might work with the map editor, it will fail in the context of a mobile application. If this is ever the case, you can simply edit the JSON file using a text editor and strip the name on the directory prefixed from the tileset section.

Let's turn this map into a game

Now that we have set up our map, we can finally start coding and re-use the assets we just created. So let's open our `app.js` file and remove any existing code in order to have a fresh start.

The scaffolding

As with every Titanium application we have developed so far, we will create a new window with a black background. We also want to interact with this window later down our code, so we will keep its reference in the `win` variable:

```
var win = Ti.UI.createWindow({ backgroundColor: 'black' });
```

This window will probably be the only Titanium control that we will be using for the whole application. From this point, all the interactions will be done using the game engine. This will allow us to have a more powerful control scheme than native controls such as `Button` and `TableViews`. On the flip side, it will require us to use a more low-level code. But don't worry, we'll go over it step-by-step.

Now that we have a window to contain our game, our first order of business will be to load the `Game Engine` module using the `require` function and keep its reference in a variable named `quicktigame2d`:

```
var quicktigame2d = require('com.googlecode.quicktigame2d');
```

We will now create our game view. It is a regular container view that will contain our game scenes and animations, and also handle touch events. Simply put, you cannot interact with the characters on the screen, but you can interact with the game view, which will in turn tell the character what to do (figuratively of course):

```
var game = quicktigame2d.createGameView();
```

We will then set our game's **frames per second** (**FPS**), which is basically the speed at which our game will run. This can only be done once and it must be done before the game loads. Since we are developing a **Role Playing Game**, `30` frames per second will suffice. Also, we will set the game view's background color to black:

```
game.fps = 30;
game.color(0, 0, 0);
```

We will now create a Game Scene object that will act as a container for all the graphical elements of the game (similar to the map, the character on screen, and any text shown):

```
var scene = quicktigame2d.createScene();
```

Loading an external asset

We will now load the map that we created earlier using the `Titanium.Filesytem` API. First, we need to get the reference to the file itself from the `assets` directory using the `getFile` function with the location and file as parameters:

```
var mapfile = Ti.Filesystem.getFile(
  Titanium.Filesystem.resourcesDirectory,
  'assets/desert.json');
```

Now that we have a reference to the file, we need to read it using the `read` function. Since the file's contents are string based, we will immediately invoke the `toString` function on the file, and store it in a variable named `file_content`:

```
var file_content = mapfile.read().toString();
```

We will then transform this string into a valid JavaScript object using the `JSON.parse` function into a variable appropriately named `mapjson`. Once this is done, we will be able to manipulate our map object just as we would manipulate any other object:

```
var mapjson = JSON.parse(file_content);
```

Now that we have our map successfully loaded into a JavaScript object, we will use its attributes to display it on the screen. Luckily, the `Game Engine` module provides us with a function that does all the heavy lifting for us.

First we need to assign the proper tileset; our map has only one tileset, so this will make it very easy. We will get the very first tileset from the `tilesets` collection and store its reference in the `desert_tileset` variable:

```
var desert_tileset = mapjson.tilesets[0];
```

We will also retrieve the ground layer using the same principle:

```
var ground_layer = mapjson.layers[0];
```

We can now create our map object using the `createMapSprite` function. This function requires a data structure as a parameter in order to determine the tileset's image location, the tile's dimensions, as well as the margin and spacing. For this, we will use the following code snippet:

```
var mapinfo = {
  image: 'assets/' + desert_tileset.image,
  tileWidth: desert_tileset.tilewidth,
  tileHeight: desert_tileset.tileheight,
  border: desert_tileset.spacing,
  margin: desert_tileset.margin
};
var map = quicktigame2d.createMapSprite(mapinfo);
```

We will then calculate the map's overall dimensions simply by multiplying the tile's dimension by the number of tiles:

```
map.width  = map.tileWidth  * ground_layer.width;
map.height = map.tileHeight * ground_layer.height;
```

The two remaining properties to be set on the map are: the tileset's global tile ID (desert_tileset in our case), and the tile's data itself. This tile data basically describes the map's layout:

```
map.firstgid = desert_tileset.firstgid;
map.tiles = ground_layer.data;
```

 Each tileset used by a map has a global ID. It allows the Game Engine module to determine which tileset is to be loaded while drawing certain parts of the map.

We will also create two constants that will be used to hold the scale factor. By default, the scale factor will be set to 1, but it might be subject to change depending on the device on which the game is running:

```
var WINDOW_SCALE_FACTOR_X = 1;
var WINDOW_SCALE_FACTOR_Y = 1;
```

We will now add our newly created map into our Game Scene:

```
scene.add(map);
```

And then push this scene into the game view:

```
game.pushScene(scene);
```

Finally, we will add our game view to our window and open the window in full screen mode. The navigation bar has been hidden so that our game occupies the entire screen:

```
win.add(game);
win.open({ fullscreen:true, navBarHidden:true });
```

Loading the game

We have created the window, loaded the Game Engine module, and created the game view, scene, map, and so on. However, if we run our application, we will only see a black screen (remember that the window we created had a black background).

The reason behind this is because we haven't started the game yet; to do this, we will add an event handler for the `onload` event from the `game` view. Inside this event handler, we will determine the screen's scale by dividing the game's width by the absolute width of the display (measured in platform-specific units), using the `Titanium.Platform.displayCaps` API:

```
game.addEventListener('onload', function(e) {
  var screenScale = game.size.width /
  Ti.Platform.displayCaps.platformWidth;
```

Now that we have determined the scale of our screen, we will adjust the game's screen size so that it will fit the display. This means that the graphics will be scaled depending on the size of the device's screen:

```
game.screen = {
  width: game.size.width / screenScale,
  height: game.size.height / screenScale
};
```

We will then update the scale factor variables that we initialized earlier with the real scale. There is a very good chance that this value will remain at 1, but recalculating this when the game starts is a more dynamic approach:

```
WINDOW_SCALE_FACTOR_X = game.screen.width  / game.size.width;
WINDOW_SCALE_FACTOR_Y = game.screen.height / game.size.height;
```

Last and foremost, we will start our game using the following code:

```
game.start();
});
```

> Note that the game screen's width and height are not set until the game is loaded. This is important to keep in mind if you are planning to do calculations based on the screen's size. That is why those are made in the event handler.

Now, why didn't we just add the `game.start()` function statement before opening the window? The reason for this is quite simple. When the window is opened, the game may not be ready to start. This is especially true for games that have lots of graphical assets to load. Therefore, the `game` view provides the `onload` event to inform us that it is now ready to start the game at our convenience. Only then can we properly start the game.

Let's fire it up!

Now that we have everything in place, we are now ready to launch our game for the first time! Though it won't be much interactive, it will give us a chance to see if all the pieces fit together. As usual, we will use the **Run** button from the **App Explorer** tab.

And there we go! We now have our first iteration for a mobile game working. There is still a lot to be done in order to make this a full-fledged game though. So let's get into it in the following sections.

Hum! Somehow it looked like a better map editor

If we closely examine our running game, we will see that the map seems to be incomplete. That's because only the ground layer has been loaded onto the map.

Let's remedy that by getting the second layer from the `layers` collection contained in our `mapjson` object. As with the ground layer, we want to store that reference in a variable for later use:

```
var items_layer = mapjson.layers[1];
```

Similar to the map editor, the items layer will be its own separate entity. Therefore, we will create a new Map Sprite dedicated to this layer using the `createMapSprite` function. And since nothing has changed in terms of tile size, borders, and similar objects, we will be able to use the same `mapinfo` variable that we have used to create the ground layer earlier in this chapter:

```
var map_items = quicktigame2d.createMapSprite(mapinfo);
```

We will also borrow the same principles as we did for the map variable for calculating dimensions, global ID, and tiles:

```
map_items.width  = map_items.tileWidth  * items_layer.width;
map_items.height = map_items.tileHeight * items_layer.height;
map_items.firstgid = desert_tileset.firstgid;
map_items.tiles = items_layer.data;
```

We will then add this new MapSprite to our Game Scene:

```
scene.add(map_items);
```

Let's see our map now

Now that we have loaded both layers from our map and also added this new layer to our Game Scene, we are ready to test our game for the second time.

Now, we have every layer from our map which gives our game a much better appearance overall. Just keep in mind that if we had more layers, we would do the same for every single layer we wanted to display.

We need a hero

Every game needs a hero and our game will be no exception to this rule. The first order of business will be to determine how he will look like on the screen. We want the hero to be able to face all the four directions when moving around the map, and we want him to have a walking animation when he is moving.

Hold that code, will you?

Just as we did with the map, we will need to have a graphical representation of our hero. Not only that, we will need a drawing for each direction to which our hero is facing as well as every frame of his walking animation.

To do this, we could use one image for every position at which our hero will appear in the game world, but a better approach will be to use a SpriteSheet. Much like a tileset, a SpriteSheet is basically an image containing all the images which we will need to display our hero at any given time.

There are a lot of free SpriteSheet that we can use. For our game, we will be using one made by a gentleman named *Sith Jester*. He has a website filled with freely usable character sheets, tilesets, objects, icons, and so on. (You can look up his website in the *Appendix, References*)

Here is what our hero's SpriteSheet will look like:

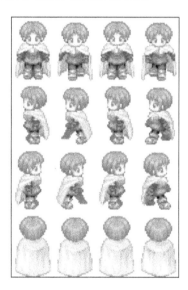

Once we have chosen our hero's appearance, we will save SpriteSheet into the assets directory under the name knight_m.png.

How does a SpriteSheet actually work?

A SpriteSheet acts as a grid where each slot contains an image of the same object (in this case, our hero). Each slot is assigned a number starting from zero at the very top left-hand corner of the grid. As we move from left to right, this number is incremented and continues to the next row once it runs out of slots. Our hero's SpriteSheet is composed of 16 images in all and has the following slot layout:

0	1	2	3
4	5	6	7
8	9	10	11
12	13	14	15

So, if we apply the grid to our real image:

- Slots 0 through 3 will show our hero going downwards
- Slots 4 through 7 will show our hero going to the left
- Slots 8 through 11 will show our hero going to the right
- Slots 12 through 15 will show our hero going upwards

As you can already see, there is a pattern here. And even though we haven't covered how our hero will move around the map yet, we can already guess how we will be interacting with the SpriteSheet depending on the direction we want our hero to go.

Bringing our hero into the scene

Now that we have what it takes to display our hero on the screen, we can start coding and bring him into our game. For this, we will create a new file in our `Resources` directory named `character.js`. This file will contain everything needed to define and interact with our hero.

Let's keep it modular

Our goal is to have a reusable `CommonJS` module that we can instantiate anywhere from within our game simply by using the `require` function. We will create a new function named `Character` that will act as our constructor function:

```
function Character() {
```

We will then load our `SpriteSheet` using the `createSpriteSheet` function. Each individual image portion representing our hero on the sheet is 32 pixels wide by 48 pixels high; there is no black border that separates each section, but we still want to have a 1 pixel margin around our hero. Knowing all these, we will pass all information to our creator function as well as the image's location. We will also store the newly created `SpriteSheet` in a variable named `self` for future use:

```
var self = quicktigame2d.createSpriteSheet({
  image: 'assets/knight_m.png',
  width: 32,
  height: 48,
  border: 0,
  margin: 1
});
```

The constructor will then return this newly created object for it to be accessible from outside the function:

```
    return self;
}
```

Since we want this module to adhere to the `CommonJS` pattern, we will export our constructor function:

```
module.exports = Character;
```

This is a basic hero implementation and admittedly it doesn't do much yet; but it will do for now. We will be adding more functionality to this class as we go along.

Putting our hero on the map

It is now time to bring our hero onto the map and display it on the screen. For this, we will load our `character` module using the `require` function and store its value in a variable named `Character`. We will then create a `hero` variable that we will be using to interact with from our code using the `Character` constructor:

```
var Character = require('character');
var hero = new Character();
```

All we need to do now, is add our hero to our Game Scene before we can take it for a spin:

```
scene.add(hero);
```

Let's see how our hero looks in the game for the very first time, by referring to the following screenshot:

Hey! Don't stay in the corner

Our very first character test went pretty smoothly but there was one thing which was odd. Our hero appeared in the upper left-hand corner of the screen, which is probably not very practical. Almost all **Role Playing Games** (RPG) position the hero right in the middle of the screen and that is what we will do too.

In the `onload` event listener, we will set the hero's x and y coordinates to half the size of the screen minus half the size of our character:

```
game.addEventListener('onload', function(e) {
  ...
  hero.x = (game.screen.width * 0.5) - (hero.width * 0.5);
  hero.y = (game.screen.height * 0.5) - (hero.height * 0.5);
```

Again, while the `Game Engine` module greatly facilitates the usage of tilesets, animations, SpriteSheets, and other graphical things, we still have to calculate our positioning. This is very different from regular applications that can rely on the layout managers to help in positioning the elements on the screen.

 Remember that we couldn't perform calculations outside the `onload` event handler since the screen's size is not initialized until this event is triggered. If we ever put it elsewhere in the code, our hero would still appear at the upper left-hand corner of the screen.

Venture around the map

We can now move on the heart of the game itself, moving our hero around the map. This section will address concepts that are pretty common in game development, such as:

- Sprites
- Game loops
- Scrolling
- Positioning
- Speed calculation

No directional pad, but that's okay

Anybody who has ever held a typical smartphone will have surely noticed that there is nothing that can act as a controller; there's no trackball and there are no directional buttons. That's not a problem; we'll just build our own.

To do this, we will use an image that will act as a controller for our game. A **Virtual Directional Control Pad (V-Pad)**, similar to the following diagram, will allow us to move our hero in every direction:

Creating our V-Pad

The V-Pad creation code is pretty straightforward; we will be using a sprite. A sprite is a two-dimensional image that can be shown on the screen. Think of it as a more powerful image view if you will.

We will create our sprite using the `createSprite` function and pass our image's full path as a parameter. Since we want our V-Pad to be translucent on screen (that way it won't hide what is underneath), we will set the `alpha` property to `0.5` (half-translucent). Finally, as with every graphical object in our game, we will add it to our Game Scene:

```
var vpad = quicktigame2d.createSprite({
  image:'assets/control_base.png'
});
vpad.alpha = 0.5;
scene.add(vpad);
```

> The `alpha` color property is a float value from 0 to 1, where 0 is completely translucent (invisible), and 1 is fully opaque.

We will then position our V-Pad centered at the bottom of the screen. Once again, we must add this code in the `onload` event handler in order for the screen's dimension to be initialized:

```
game.addEventListener('onload', function(e) {
  ...
  vpad.x = (game.screen.width * 0.5) - (vpad.width * 0.5);
  vpad.y = game.screen.height - vpad.height;
```

Is someone touching the V-Pad?

From this point, everything is in place to start interacting with our V-Pad. There are two things we need to capture throughout the lifetime of our game in order to keep the interaction responsive: is the user touching the controller (isVpadActive)? And if so, where is his finger precisely, so that we can determine the directions, touchX and touchY, in which to move our hero. Therefore, we will declare the following three variables so that we can access this information at any given time:

```
var isVpadActive = false;
var touchX, touchY;
```

Since we cannot listen to the touch events only on the controller part of the screen, (remember these are not Titanium views), we will need to add event handlers on the game view, and then determine if a touch event is within the bound of the controller:

Our first EventListener will be on the touchstart event:

```
game.addEventListener('touchstart', function(e) {
```

We will then assign our touchX and touchY variables by multiplying the event coordinates by the scale factor we calculated earlier. This is done because all our graphical elements are positioned based on the same screen factor. Therefore, we must make sure that all the interactions are done on the same base:

```
touchX = (e.x * WINDOW_SCALE_FACTOR_X);
touchY = (e.y * WINDOW_SCALE_FACTOR_Y);
```

We will then determine if the user touched the V-Pad or not by using the sprite's contains function. This is a very powerful function which basically returns true if the screen coordinates passed as parameters are included in the space occupied by the sprite. If not, then it returns false. Once this is determined, we assign this value to the isVpadActive variable so that we can take action depending on its value:

```
isVpadActive = vpad.contains(touchX, touchY);
});
```

We also do exactly the same thing for the touchmove event. This event listener is also very important because we need to know if the user is moving his or her finger on the V-Pad to change the hero's direction or speed:

```
game.addEventListener('touchmove', function(e) {
  touchX = (e.x * WINDOW_SCALE_FACTOR_X);
  touchY = (e.y * WINDOW_SCALE_FACTOR_Y);
  isVpadActive = vpad.contains(touchX, touchY);
});
```

Once the user lifts his or her finger off the screen, there is no need to do any calculations anymore. We will then set the `isVpadActive` variable to `false`:

```
game.addEventListener('touchend', function(e) {
  isVpadActive = false;
});
```

Giving some visual feedback to the user

If we run the game right now, it would load perfectly; but it wouldn't do anything even if you touched the screen. Looking at the code we just added, it does not make any sense when you think about it. All we did was assign a variable depending on if the user was touching the controller or not. Well, let's do something with this variable to give our users some visual feedback when he or she touches the V-Pad.

We will create a new function called `updateVpad` that will change the controller's background color depending on whether it is active or not:

```
function updateVpad() {
  if (isVpadActive) {
    vpad.color(0.78, 0.78, 0.78);
  } else {
    vpad.color(1, 1, 1);
  }
}
```

Aren't we forgetting something?

Touch events are fired every time there is a slight change in the finger's position on the screen. So if the `updateVpad` function was to be called on every event, it would be called thousands of times, thus having a significant performance impact.

To address that, we will call our function at a fixed interval using the `setInterval` JavaScript function. This function is very simple; it will wait for a specified number of milliseconds, and then execute a specified function every time this interval is reached until it is explicitly stopped in the code.

We will create a global variable for our interval timer so that we can stop it later on:

```
var updateVpadTimerID = 0;
```

We will call our `updateVpad` function every 66 milliseconds (arbitrary value), in order to have a very quick reaction time when the user interacts with the game. We will call this function right after the game is started in the `onload` event listener. We will also save the reference to this timer in the `updateVpadTimerID` variable for later use:

```
game.addEventListener('onload', function(e) {
  // All the dimensioning and positioning code...
  game.start();
  updateVpadTimerID = setInterval(function(e) {
    updateVpad();
  }, 66);
});
```

We will run the game again to see that the V-Pad changes its color when we touch it, but there is no interaction if we touch any other area of the screen.

Moving our hero around

Now that we have a working virtual controller in place, we can start using it to move our hero around the map. All the code related to character movement will be located in the `updateVpad` function when the `isVpadActive` variable is a true condition.

So all of the code from this subsection will have to be inserted at the following location:

```
function updateVpad() {
  if (isVpadActive) {
    // Insert Movement Code Here
```

Seeing the future

One thing about game development in general, is that you never actually move the object itself until you are 100 percent sure whether you are going to move it in the next frame. This is done to avoid doing calculations as you move the graphic elements on the screen, which can degrade your performance. So, what we want to do is determine where all our elements will go and keep their future coordinates in the game's memory.

At first we will determine the speed at which our hero is moving on the map. To achieve this, the formula is a lot simpler than it appear when we first look at the code. Basically, we calculate the distance between the absolute center of the V-Pad and where the user touched on the controller. The farther away from the center, the faster the hero will walk. Once the vertical and horizontal speeds are determined, we will store those values into variables to use when the moment is right. (This is not yet the moment to move anything on screen.)

```
var speedX = (touchX - (vpad.x + (vpad.width * 0.5))) * 0.2;
var speedY = (touchY - (vpad.y + (vpad.height * 0.5))) * 0.2;
```

We will then determine what will be the next screen coordinates of our hero, as well as the map and keep those values for the variables:

```
var nextHeroX = hero.x + speedX;
var nextHeroY = hero.y + speedY;
var nextMapX = map.x - speedX;
var nextMapY = map.y - speedY;
```

Living the future

It is now time to move all our graphical elements on the screen in a single operation, since everything has already been calculated. Once again, these conditions look a lot more complicated than they actually are. We will cover them one by one.

If we want to move our hero on the horizontal axis and the future position is lower than the screen's width, we will update the hero's position:

```
if (nextHeroX > 0 && nextHeroX < (game.screen.width-hero.width)) {
    hero.x = nextHeroX;
```

Also, we will check if the map's next horizontal position is a negative number and still higher than what is left on the map that is not displayed on the screen, then we will scroll the map as well as the map items (remember that there are two layers on this map). Of course, if there is no more map to be scrolled, like when the hero reaches the end of the map, no action is taken. This will give the user the sensation that our hero has hit some sort of invisible barrier:

```
} else if (nextMapX <= 0
    && nextMapX > (-map.width + game.screen.width)) {
    map.x = nextMapX;
    map_items.x = map.x;
}
```

We will then apply exactly the same algorithm to the vertical positions in order to have similar behavior:

```
if (nextHeroY > 0
    && nextHeroY < (game.screen.height - hero.height)) {
  hero.y = nextHeroY;
} else if (nextMapY <= 0
    && nextMapY > (-map.height + game.screen.height)) {
  map.y = nextMapY;
  map_items.y = map.y;
}
```

Is it a game yet?

The short answer to this question would be: "Yes, it is a game and we can start playing with it." And this is precisely what we will do by running it right now.

Our game looks similar to the following screenshot:

We can now move our hero around the map using the onscreen V-Pad and we can see the map scrolling when our hero reaches the edge of the screen.

Our hero moves, but he's still pretty stiff

One thing you may have noticed while moving our hero around the map is that no matter where our hero goes, he always faces the same direction (downwards). While this behavior was widely accepted in the 1980s, the same cannot be said for today's games, even for mobile games. We will now address this in the following section.

Make sure he walks in the right direction

As the title says, there are two new things our hero must be able to do. The first one is walking, which will involve animation. The second one is facing the right direction, which will involve using more features of our sprite sheet.

Since our `Character` class contains everything related to our hero, we will open the `character.js` file and add a new function named `turnTowards` to the `self` variable. Since our constructor function returns the `self` variable, every single character in our game will now have a `turnTowards` function, hence making it accessible to the outside world.

The character's new direction will be passed to the function:

```
self.turnTowards = function(newDirection) {
```

Before we do anything graphically related, we will check to see if the new direction is different from the one our character is already facing. If it is the same, no action is needed.

```
if (self.direction !== newDirection) {
```

We will then show a different animation depending on the direction the character is facing. This takes us back where we defined our `SpriteSheet` and each row represented our hero walking towards one direction. Each row had four images and we wanted to animate the four images that were to the right when the character moved.

```
switch (newDirection) {
  case "DOWN":
    self.animate(0, 4, 250, -1);
    break;
  case "LEFT":
    self.animate(4, 4, 250, -1);
    break;
  case "RIGHT":
    self.animate(8, 4, 250, -1);
    break;
```

```
    case "UP":
      self.animate(13, 4, 250, -1);
      break;
    default:
      self.animate(0, 4, 250, -1);
  }
```

Once our character is engaged in the new direction, we update the object's current direction:

```
    self.direction = newDirection;
  }
}
```

The `animate` function takes four parameters:

- Which will be the first frame of animation from the SpriteSheet?
- How many frames will the animation have? (Starting with the one defined in the first parameter)
- How many milliseconds between each frame?
- How many times will the animation loop be played? (-1 makes it infinite)

Now that our character class is capable of facing four directions as well as walking in an animated manner, we will update our `updateVpad` function (the one that moves our hero on the screen). Back to our `app.js` file, we will create a new variable that will be used to hold our hero's current direction. Since we want to be able to see the hero's face when the game is loaded for first time, we will set the default value to DOWN:

```
var heroDirection = "DOWN";
```

We will now determine the direction to which our hero must face depending on the direction pressed on the V-Pad. If the horizontal speed is higher than the vertical speed, it means our hero will be facing either towards the left or right. We will then check if the number is positive or negative. If it is positive, our hero will face to the right, and if not, it will face to the left:

```
function updateVpad() {
  if (isVpadActive) {
    ...
    if (Math.abs(speedX) > Math.abs(speedY)) { // Horizontal
      heroDirection = (speedX < 0) ? "LEFT" : "RIGHT";
```

The principle is exactly the same for the vertical axis, as you saw in the preceding code.

```
} else { // Vertical
    heroDirection = (speedY < 0) ? "UP" : "DOWN";
}
```

Once the direction is determined, we will call the `turnTowards` function and pass the new direction as a parameter:

```
hero.turnTowards(heroDirection);
```

With this in place, we can now test our game once more to see our animated hero walking in different directions.

Make sure he stops in the right direction

Now that we have our hero walking, we can't seem to make it stop. This is due to the `-1` parameter that we passed to the `animate` function earlier, which started the animations in an infinite loop. But this was necessary since we wanted our hero to walk as long as our user pressed the controller pad. This is not an insurmountable problem, since all we have to do is to stop the animation when the user releases the V-Pad.

To achieve this, we will add a new function aptly named `halt` to our `Character` class. So going to our `character.js` file, we will add this new function to our constructor function. The function checks the direction in which the character is facing and simply pauses the animation at the appropriate slot from our `SpriteSheet`:

```
self.halt = function() {
  switch (self.direction) {
  case "DOWN":
    self.pauseAt(0);
    break;
  case "LEFT":
    self.pauseAt(4);
    break;
  case "RIGHT":
    self.pauseAt(8);
    break;
  case "UP":
    self.pauseAt(13);
    break;
  default:
    self.pauseAt(0);
  }
};
```

Now once again, we need to call this newly created function when the user releases the V-Pad. So going back into our `app.js` file, we will call the `halt` function when the `isVpadActive` variable is `false`:

```
function updateVpad() {
  if (isVpadActive) {
    // Our hero moves code...
  } else {
    ...
    hero.halt();
  }
}
```

Its now time for one of our final test runs to see that our hero animates itself walking from point A to point B, but stops when the V-Pad is released (both while facing the right direction, we might add).

Putting the finishing touch

While our game has all the features that we wanted it to have, there are still little things we can add in order to give our game a better appeal. While not mandatory, these improvements are essential when developing world-class applications. Let's go over them quickly.

The user is touching the V-Pad, but where exactly?

While it is true that we give some visual feedback when the user touches the game controller, there is no way for our user to know where exactly he or she pressed the controller. To achieve this, we will be adding a visual indicator that basically appears at the exact coordinates of the touch events.

For this, we will use a small, transparent image with a small particle effect on it, so that it is still somewhat translucent when displayed. We will create a new sprite using the `createSprite` function and by passing our image path as a parameter:

```
var vpad_nav = quicktigame2d.createSprite({
    image:'assets/particle.png' });
```

Since we want this image to only be shown on demand, we will immediately hide it so that it is not seen anywhere. We will also give a yellow color to our sprite so it will stand out well under our fingers:

```
vpad_nav.hide();
vpad_nav.color(1, 1,  0);
```

As with every graphical item from our game, we will add it to our Game Scene:

```
scene.add(vpad_nav);
```

Since similar interactions take place in our new sprite and the V-Pad, we will modify the `updateVpad` function again. When `isVpasActive` will be `true`, we will position the sprite right under the touch event's coordinates. It will also be centered so that it doesn't look offset from where it is supposed to be on the screen. Finally, we will invoke the sprite's `show` function so that it becomes visible on the screen.

Again, the opposite is true for when `isVpasActive` will be false; we will simply hide out sprite so that it doesn't stay on screen one second more than necessary:

```
function updateVpad() {
  if (isVpadActive) {
    vpad_nav.x = touchX - (vpad_nav.width  * 0.5);
    vpad_nav.y = touchY - (vpad_nav.height * 0.5);
    vpad_nav.show();
    ...
  } else {
    ...
    vpad_nav.hide();
  }
}
```

Be sure nothing overlaps

Depending on the order in which we added our graphical elements to the Game Scene, there might be things that overlap each other, or even worse, some graphical elements may be covered entirely (rendering them invisible). To address these concerns, we will set each element's Z-order property. This Z-order property is similar to Titanium's `zIndex` property and is indicated by the property `z`.

At the very end of our code—usually before the window opens—we will add the following code:

```
map.z = 0;
map_items.z = 1;
hero.z = 2;
vpad.z = 3;
vpad_nav.z = 4;
```

When you take the time to read it, it makes sense. First we add the map, then we add the map's items on top of the map, and then we want our hero over the map because it makes no sense for him walking under the rocks and sand. Once the game is in place, we want our V-Pad to sit on top of everything related to game so that it doesn't interfere with the controls. Lastly, we add the yellow dot navigator that will sit atop the controller.

 This might not be necessary for our game if we have been careful when adding our elements, but this step becomes a necessity when dealing with huge amounts of graphical assets (for example, a shooting game).

Cleaning up after ourselves

When a user leaves an application, the application is put into the background, put to sleep after that, and after a period of time, it is completely stopped. But there are differences, depending on which platform our game is running. For instance, when an Android application is stopped, the ongoing timer we started earlier will not be stopped automatically. So it is our responsibility to stop it through our code.

We will add an event listener on our main window's `android:back` event. This, as its name states, handles the back button on every Android device. We will call the generic JavaScript `clearInterval` function by passing our timer variable that we initialized when we started the timer. Finally, since we want to quit the application by clicking on the back button, we will close the window:

```
window.addEventListener('android:back', function(e) {
    clearInterval(updateVpadTimerID);
    window.close();
});
```

 While this code will be run only on Android devices, we do not need to put it into an `if (isAndroid)` code block, since it will simply be ignored under iOS.

Summary

In this chapter, we covered the development of a very basic JRPG that allows our player to walk around the map and all the steps to achieve it.

First, we saw that it takes a lot of tools to build a game; code is not enough in this case. We learned how to incorporate a native game engine into a Titanium application in order to enjoy far more graphical performance than regular views.

We covered the uses of external tool to build our game assets, such as the map editor. We went over concepts such as tilesets, layers, individual tiles, and so on. We also learned how to manipulate graphics with concepts such as sprites, SpriteSheets, and animations.

More generally speaking, we covered concepts such as the gaming loop and event interaction. We also covered finer-tuned touch events such as `touchstart`, `touchmove`, and `touchend` so that they can work when the user drags his or her finger on the screen. Such concepts are used to some extent even when developing a bigger game.

With all that has been learned in this chapter, we developed a very basic JRPG that is fully-functional and that can serve as a base for other, more complex, game projects.

In the next chapter, we will bring along friends on our epic adventure by adding online capability to our game.

7
JRPG – Second to Last Fantasy Online

While most of the mobile games can be played alone, there are a lot of cases where the experience becomes more enjoyable when the game is a multiplayer one. This chapter will walk us through the process of turning our JRPG into a multiplayer game. We will create a rudimentary game server where players can connect to and interact with each other. From there, we will iterate on the existing game and bring a multiplayer component to it. We will allow multiple players to roam around the map and interact with each other through a basic chat feature. In order to differentiate each player we will add the ability to change the hero's appearance.

This chapter differs from all the other chapters of this book, since it relies on the codebase covered in *Chapter 6, JRPG – Second to Last Fantasy*. While reading the previous chapter is strongly recommended in order to understand the changes covered, it is not 100 percent mandatory. This is also a much longer chapter that covers a lot of cool concepts. We will go through each concept in an orderly manner so that it makes sense down the road.

By the end of this chapter, we will have a fully-functional multiplayer game and covered the following concepts:

- Understanding how a game server behaves
- Interacting with the game server using WebSockets
- Sending information to the game server
- Applying other players' interactions to the current game
- Using the `ScrollableView` component
- Implementing a basic chat feature

Creating our project

As every application we created so far, we will create our new project by selecting the **File** | **New** | **Mobile Project** menu from Titanium Studio and fill out the wizard forms with the following information:

Field	Value to be entered
Project Template	Default Project
Project Name	SecondToLastFantasyOnline
Location	You can either: • Create the project in your current workspace directory by checking the **Use Default Location** checkbox • Create the project in a location of your choice
App Id	com.packtpub.hotshot.secondtolastfantasy.online
Company/Personal URL	http://www.packtpub.com
Titanium SDK Version	By default, the wizard will select the latest version of the Titanium SDK. This is recommended (while writing this book, we were using Version 3.1.3.GA)
Deployment Targets	Check **iPhone**, **iPad**, and **Android**
Cloud Settings	Uncheck the **Cloud-enable this application** checkbox

Project creation is covered more extensively in *Chapter 1*, *Stopwatch (with Lap Counter)*. So feel free to refer to this section if you want more information regarding project creation.

Recycling saves our time

Since we don't want to redo all the work that was done, we will copy all the files from the codebase in *Chapter 6*, *JRPG – Second to Last Fantasy*, into our newly created project. More explicitly, we will copy every single JavaScript file (as well as all graphical assets) from the Resources directory into our newly created project. But most importantly, we won't copy the tiapp.xml file from the previous project since it is specific for each application.

You might get some warnings, since there are files that share similar names between the two projects. Simply overwrite all the files in order to have a project that is an exact replica of the previous one (but do not delete the `tiapp.xml` file, of course).

 For readers who didn't go through the previous chapter, you can easily download the source code from the public GitHub repository (`http://bit.ly/14SHQ39`).

Before we get our hands dirty

As with all our previous applications, we need to have a clear vision of what we need to accomplish before digging into the code. Our multiplayer game will rely on a central game server and every game client will connect to this same server. Communication between the two will be bidirectional, since one player's actions must be perpetuated to all the other players, creating a richer experience for everybody.

Some things to be considered

Online multiplayer games usually transmit and receive quite a lot of data. But on mobile devices, this can quickly become a problem since we can never be sure of the reception quality. A game might perform well with Wi-Fi, but crawl to a stop over cellular data connections. Furthermore, cellular networks are much less reliable depending on where the user is located. This is an important mobile principle. Many new mobile developers sometimes learn this the hard way.

Such use cases and their limitations require us to be mindful about the amount of data and the frequency at which it is being transmitted.

If we want to develop a standard web application using AJAX, we will use the following principles:

- The client sends the actions done by the player to the server at a regular interval
- The server receives those actions, processes them, and sends back all of the actions from the other players to the client
- The client receives all the other players' actions from the server and applies them to the screen

- Rinse and repeat

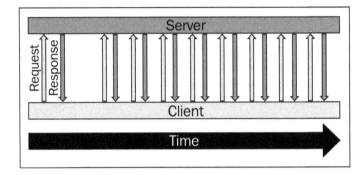

While such an approach is sustainable on current browsers running on machines with huge quantities of RAM, we can already see that the same cannot be applied on mobile devices.

WebSockets

Making many HTTP requests at high frequency can create a bottleneck over slower connections. Even if there is no player action, there is still the overhead for the request itself and that can add up over time. A more effective approach would be for the server and the client to exchange information on the required basis.

There is a mechanism called WebSockets that allows us to do just that. Fundamentally, WebSockets use long-lived HTTP connections to reduce the latency with which messages are passed to the server. This is a drastic improvement over the previous approach since it reduces the amount of information being passed over the network, as shown in the following diagram:

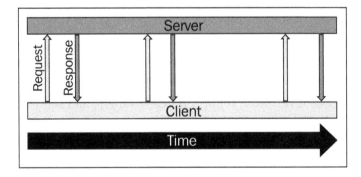

In order for WebSockets to work though, it is imperative that not only the client, but also the server supports WebSockets. Since iOS and Android already support this feature under the hood, we can be sure that our game will be able to have the same behavior on every platform.

WebSockets is a generic term that designates what we just covered in this section. There are many implementations for almost every language out there. So, we are bound to see more and more sites relying on such mechanisms to improve performance as well as reduce the load on their servers.

Setting up things

Before we start writing any code, we need to put some things in place in order to have a proper foundation on which to rely on.

The server side

Every online game requires a game server of some kind and our game will be no exception. It will be used to keep tabs on all the connected players and will relay all the games' interactions to the players.

Our game server must meet the following requirements:

- It must be accessible through a standardized protocol (HTTP in our case)
- It should have Web Socket support
- It should be easy to develop (we are the ones coding it after all)

There are a lot of web servers out there that can meet these requirements. All these requirements are based on different technologies under the hood, requiring us to master another language in order to use them. Also, we would need to have a conversion process in place in order to pass data between the server and the client. JavaScript can parse JSON and interpret them as native objects, but that isn't necessarily the case for other languages we find on most of the web servers, such as C, PHP, or even Java.

Enter Node.js

Node.js is different from most web servers as it is entirely built on the V8 JavaScript runtime, which was developed by Google for their Chrome web browser. Node.js is incredibly fast and has a very active community, so there are many modules out there to achieve almost everything you would want to do. But probably the biggest benefit of using Node.js is that it is actually written in JavaScript; a very dynamic language that makes it easy to develop event-based code.

This chapter won't go into all the inner workings of Node.js, but will cover everything we need to get our small server set up. Just keep in mind that Node.js is a framework that provides an easy way to build scalable network programs (such as web servers), for specific purposes.

Installation

The very first thing we need to do is to determine if Node.js is installed on our development machine. Usually, installing Titanium Studio takes care of that for us.

There is a very simple test we can run to check if Node is installed and correctly configured. Simply type the following command from the command line:

```
$ node --version
```

```
v0.10.13
```

If Node.js is properly configured, the command will return the server version. While writing this book, we used Version 0.10.13.

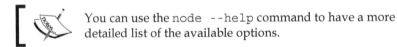 You can use the node --help command to have a more detailed list of the available options.

If the command does not return the version, you need to install the server.

You can either:

- Go to Titanium Studio and select the **Help | Check for Updates** menu. This should detect if the Node.js installation is missing and provide you with the option to automatically do so.
- If that doesn't work, download the appropriate version from the official Node.js website (http://www.nodejs.org) and follow the instruction from the home page.

Creating our server project

Now that we have our Node server installed, we need to create a working directory for our server project. This is a completely separate directory from the one we created earlier for our mobile application, since this project is targeted to the server. Unlike every mobile application project we created so far, there are no wizards or configurations involved here.

We will create a directory named `SecondToLastFantasyOnlineServer` and then go to this newly created directory using the following command:

```
$ mkdir SecondToLastFantasyOnlineServer
$ cd SecondToLastFantasyOnlineServer
```

 The location of the directory can differ on your machine depending on your configuration. But it is considered as a "good practice" to have it located at the same level as the game client directory.

socket.io

There are many different WebSockets implementations out there, but probably the most widely used would be socket.io. It is entirely JavaScript-based and is compatible with a desktop as well as a mobile browser. What it does under the hood is pretty clever actually; if a desktop browser has an Adobe Flash player enabled, for example, it will leverage that to increase performance. If not, it will use another channel to achieve its goal. Such an approach allows us to focus on the code without having to worry about which platform will be used at the end.

Also, socket.io can be used on the server, allowing us to use the same code on both the ends (this was one of the main reasons that made us choose Node.js).

Installing a JavaScript module on our server sounds complicated

We now install the socket.io module into our project instance. We can always download it directly from the GitHub repository, create an appropriate directory structure, and copy the files to the right location. But we would then run into version and dependency problems, not to mention that we would have to do this all over again every time the module needs to be updated.

Luckily, there is a neat utility program that does all the heavy lifting for us, called **NPM** (**Node package manager**). It will allow us to install, update, and even uninstall Node.js modules as well as all the dependencies. When called, npm will query an online repository and automatically fetch the necessary files needed for this module.

We will install the Node.js module called socket.io into our project directory using the following command:

```
$ npm install socket.io
```

 Your current user might not have sufficient rights to install things using the npm command. To resolve around this issue, simply prefix your command with sudo (superuser do) and you should be fine.

Once the command has successfully run, we will have a new directory called node_modules in our project. This is where all the downloaded files are located. And by convention, when we reference the socket.io module in our code, Node.js will go into this same directory to access the module's code.

Making sure everything is in place

Now that our module has been downloaded and installed, we need to make sure that it is correctly configured. To achieve this, we will call the npm command with the list switch. This will list all the modules installed as well as their dependencies for this project location.

Run the following command:

```
$ npm list
```

The output of the preceding command is as follows:

```
/SecondToLastFantasyOnlineServer
└─┬ socket.io@0.9.14
  ├── base64id@0.1.0
  ├── policyfile@0.0.4
  ├── redis@0.7.3
  └─┬ socket.io-client@0.9.11
    ├─┬ active-x-obfuscator@0.0.1
    │ └── zeparser@0.0.5
    ├── uglify-js@1.2.5
    ├─┬ ws@0.4.25
```

```
|   ├── commander@0.6.1
|   ├── options@0.0.5
|   └── tinycolor@0.0.1
└── xmlhttprequest@1.4.2
```

The client side

Now that we have socket.io installed on our game server, we need to add something similar into our mobile application. If this were a normal HTML page, there wouldn't have been any issues, since the server would serve the necessary files and the client would just use them with a simple HTML `<script>` tag. But mobile applications cannot download any code in order to run it later. This makes sense since there wouldn't be any way of preventing a malicious application from downloading and executing code. Apple and Google have made this an integral part of their terms and conditions, and the applications that do not abide by this rule are rejected.

Knowing this, we need to embed a socket.io client into our application to interact with our server. Since this functionality isn't supported out of the box by Titanium, we shall rely on an extension module.

The Web Socket module

As we've already done in *Chapter 4, Interactive E-Book for iPad* and *Chapter 6, JRPG – Second to Last Fantasy*, we will use a native module to add the Web Sockets functionality to our Titanium application. We will be using the **TiWS** (**Titanium WebSockets**) module since it supports quite a few implementations of WebSockets (socket.io is the one that we are interested in). It is also available on iOS and Android and is open source and is under the Apache License, Version 2.0. It was developed by a gentleman named *Jordi Domenech*.

Downloading the module

The module can be obtained via the Appcelerator Marketplace at the locations shown in the following table. For those of you who would like to have a look at the code (or even build it yourself), you can do so by accessing the following repository:

File description	Download link
TiWS module 0.3 for iOS (from the Appcelerator Marketplace)	`http://bit.ly/1529TwA`
TiWS module 0.1 for Android (from the Appcelerator Marketplace)	`http://bit.ly/142HLKG`
The module's source code repository	`http://bit.ly/L8eFzj`

Notice that the iOS and Android implementations don't share the same version numbers. We will address that in the following section.

Installing the module

Now that we have retrieved our module, we need to add it into our project. Just as we did in *Chapter 4, Interactive E-Book for iPad*, and *Chapter 6, JRPG – Second to Last Fantasy*, we will copy the module's archive file (`downloads_2825_net.iamyellow.tiws-iphone-0.3.zip` for iOS or `downloads_3158_net.iamyellow.tiws-android-0.1.zip` for Android) into our `/Resources` directory and then delegate the installation process to Titanium by selecting the **Run** menu.

Now that the module is installed, we will configure it by updating the `<modules>` section of our `tiapp.xml` file. We will update and simply add the following lines identifying the module, its platform, as well as its version.

If we were planning to develop an iOS-only game, we would add the following reference:

```
<modules>
  <module platform="iphone" version="1.2">
    com.googlecode.quicktigame2d
  </module>
  <module platform="iphone" version="0.3">
    net.iamyellow.tiws
  </module>
```

For an Android-only game, we would be using the following XML code. Notice that the `platform` and `version` attributes are different depending on the target platform:

```
<modules>
  <module platform="android" version="1.2">
    com.googlecode.quicktigame2d
  </module>
  <module platform="android" version="0.1">
    net.iamyellow.tiws
  </module>
</modules>
```

Since our game is already playable on both iOS and Android, we want to keep things in the same way. Therefore, we will be using the following XML snippet for our `modules` declaration:

```
<modules>
  <module platform="iphone" version="1.2">
    com.googlecode.quicktigame2d
  </module>
  <module platform="android" version="1.2">
    com.googlecode.quicktigame2d
  </module>
  <module platform="iphone" version="0.3">
    net.iamyellow.tiws
  </module>
  <module platform="android" version="0.1">
    net.iamyellow.tiws
  </module>
</modules>
```

As usual, we will validate the correct configuration of the modules by looking at the **Modules** section from the **Overview** tab in Titanium Studio. As we can see in the following screenshot, both the modules are configured, each with their respective version depending on the target platform:

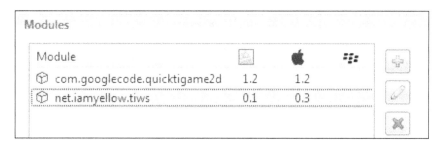

This will work, but at the same time, it won't

While the module is properly installed and configured, there is still one last operation that we need to do. As we mentioned earlier, the TiWS module supports many different Web Socket implementations. In order for our module to work properly with socket.io, we need to copy the `socket.io.js` file from the `example` directory into our application's `Resources` directory.

 You will find a similar file in the Android version of the module. Both files are identical, so you can use either without having to worry about any compatibility issues.

Some of you might ask: "Why don't we just use the same `socket.io.js` file from the server? This way, we will be sure that both of the versions are compatible." This version of the `socket.io.js` file is an updated one used in order to work inside a Titanium application properly. Using the same file as the server just won't work.

Coding our game server

Now that our Node.js server is properly installed and our socket.io module is configured on both the server and the client, we can finally start coding. We will first go over the server code, cover all the games' interactions, and then move onto the client side and interact with the said server.

Creating a web server in less than 10 lines of code

One of the great things about Node.js is that it uses the familiar JavaScript language. It also provides us with a very powerful set of APIs to achieve our goals. We will start by creating our web server in order for our players to connect to it.

First, we will create a new file named `app.js` in our server project. Inside this file, we will code our entire game server. First we will load the `http` module using the `require` function and we will store its reference in a variable named `http` as follows:

```
var http = require('http');
```

Next, we will create our HTTP server using the following code and we will store its reference in the `server` variable for later use:

```
var server = http.createServer();
```

Now that our server is created, we will add an event handler on the `request` event every time someone queries the server. Just as we did with a Titanium event listener, the second parameter expects a callback function that will run when the event is triggered. In this particular case, the server will always return the HTTP status code `200` (OK) as well as a string mentioning that the server is running:

```
server.on('request', function(req, res) {
  res.writeHead(200, { 'Content-type': 'text/plain' });
  res.end('Server is running...');
});
```

Usually, we should expect a lot more code logic in such a function (different use cases, error management, and so on). But this will do for our test case. Keep in mind that all HTTP interactions will be done through socket.io and not this function.

The last thing we need to do is to inform the server to listen on a certain port. We will pass a callback function that will simply output a message to the console when the server is started:

```
server.listen(8080, function() {
  console.log('Listening at: http://localhost:8080');
});
```

We now have a basic (yet functional) web server and we managed to do this by writing less than 10 lines of code.

Taking our server for a spin

Let's run our newly created web server and see if it works. From the command line, we will launch our server using the `node` command and pass it the name of the file to be executed.

The following is the output when we run the command:

```
$ node app.js
Listening at: http://localhost:8080
```

Okay, so we at least know that it starts without a hitch. But does it actually handle HTTP requests? The easiest way to verify that is simply by pointing our web bowser to the server's URL (not forgetting to specify the port number) as follows:

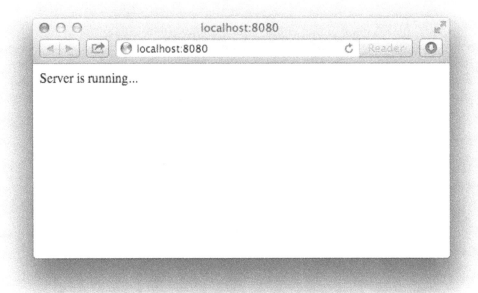

Now that we have verified that our web server works, we can move on and turn it into an actual game server. To stop the running server, simply press the *Ctrl + C* key combination.

Note that Node.js does not automatically reload the JavaScript files when it is running. Meaning that if you ever modify a source file, you will need to stop the server and restart it again in order for your changes to be loaded. There are of course modules, such as RequireJS (https://github.com/jrburke/requirejs), that offer such functionality, but this is beyond the scope of this chapter.

Keeping tabs on our players

We need our server to keep tabs on all the connected players. Since this is a fairly simple server, we will store all the connected players in a global array. Since this will be a rudimentary server, this array will not be persistent when the server restarts. We create the array as follows:

```
var players = [];
```

This array will basically contain `Character` objects created by the game client. Since we are using JavaScript on both the ends, we can pass actual objects instead of having to convert our objects every time we need to serialize them. This is a very powerful feature that Node.js has to offer.

Serialization is the process of translating an object into a format that can be stored (using XML, JSON or any another format), to a file, memory, or even transmitted over a network, and resurrected later on the same or a different machine.

Creating our socket.io connection

Now that we have our web server, we will reference the `socket.io` module and assign it to the variable aptly named `socketio`, as seen in the following code:

```
var socketio = require('socket.io');
```

Once we have a hold on the socket.io module, we can instruct it to listen to the server's connections. From there, (in the same statement) we will add an event handler for the `connection` event, which means that every time a new client connects, the callback function will be called and will have the web socket's reference as the parameter:

```
socketio.listen(server).on('connection', function (socket) {
  // Insert socket code here!
}
```

With this `socket` object, we will be able to add event handlers only for this connection, and everything that is done in this function will be exclusive between the server and the client. There are, of course, ways in which you can communicate with the other clients if needed (for example, when you move, you want to inform others about your new position). We will get to those shortly.

It is imperative that all the game actions be located inside this function, since all the code contained in this function represents a single web socket connection with the client. If we were to put code outside of that function, it wouldn't work because we wouldn't have any socket object to interact with it.

Game interactions

The first order of business in a multiplayer game is that our server needs to be informed who is playing, which directly translates to: "Who is in our `players` array?"

In order to do that, we will simply create custom events on our `socket` object when a player joins or quits the game.

A player joins the game

Although we have a connection with the client socket (defined earlier), that doesn't mean the player will automatically join the game. Therefore, we will create an event listener for the `join` event on our `socket` object. This handler will have a callback function that will have the `newPlayer` object as the parameter:

```
socket.on('join', function (newPlayer) {
```

The first thing we will do when the player connects to it is send him the list of all the players already connected to the server. These players don't exist on his local game yet, so he needs to retrieve them. To do this, we will simply loop through the `players` array and fire a `playerjoined` event on the client's socket using the `socket.emit` function. Each call passes the actual JavaScript object to the event. There is no need for `JSON.stringify` or some other translation process here:

```
for (var i=0, len=players.length; i < len; i++) {
  socket.emit('playerjoined', leanify(players[i]));
}
```

The next step is to add the new player into the `players` array. We do a simple check if the player isn't already present in the array to avoid duplicates (connection issues, incoming call, and so on). If the player isn't present already in the server, we simply add it to the `players` array as follows:

```
players.contains(newPlayer, function(found) {
  if (!found) {
    players.push(leanify(newPlayer));
```

Once the new player is added to the game, we need to notify every other player that a new player has joined the game. We will fire the same `playerjoined` event, but this time using the `socket.broadcast.emit` function. This function is quite unique since it will be broadcast to all the connected players except the player that has the socket connection (in this case, the new player):

```
            socket.broadcast.emit('playerjoined', leanify(newPlayer));
        }
    });
});
```

A player quits the game

When a player quits the game, we will do the same thing we just did but in reverse. So we will add an event handler to the `quit` event. The `callback` function will have the player who just quit as a parameter:

```
    socket.on('quit', function (player) {
```

We will then loop through every connected player on the server until one of them has the same unique ID as the quitting player's unique ID. Each device has a globally unique ID (UUID); this way we can be sure that the game server knows every single player uniquely. We will then remove the object from the array as follows:

```
        for (p in players) {
          if (players[p].id === player.id) {
            players.splice(p);
```

We also need to inform all of the other players still connected that this player has quit the game, once again using `socket.broadcast.emit` with the quitting player as a parameter:

```
            socket.broadcast.emit('playerquit', leanify(player));
          }
        }
    });
```

JavaScript arrays don't have a contains function?

Some of you might have noticed that we have been using a `contains` function to determine if a player was already inside the `players` array. Although such a function might exist in other languages, this is not the case with JavaScript. But such a limitation can be overcome by extending the `Array` class using the `prototype` property. This will add this extra function to all the `Array` instances:

```javascript
Array.prototype.contains = function(k, callback) {
  var self = this;
  return (function check(i) {
    if (i >= self.length) {
      return callback(false);
    }

    if (self[i].id === k.id) {
      return callback(true);
    }

    return process.nextTick(check.bind(null, i+1));
  }(0));
}
```

This is not your typical `contains` function where we simply loop through every item and return `true` if the item match is found. Although we will not go into all the details of this function, it is important to remember that Node.js is single threaded, meaning that accessing the array while it is being modified can create deadlocks. Therefore, this implementation uses specific node APIs that would prevent such a risk.

Player interactions

Now that the basic game management functions are implemented, we will implement the player-related events. The basic idea remains pretty similar to the previous events. Every time a player takes an action in his game, it is the server's duty to broadcast these same actions to the other players.

A player moved around the map

Our first version of the game gave us the ability to roam our hero around the map. What we want to do here is to notify every other player of our hero's current position on the map.

We will create an event listener for the move event and it will take the moving player as the parameter for the callback function. The server doesn't keep the current position of the connected players (remember that this is a basic server). So every time a player moves, we broadcast the event to all of the other players:

```
socket.on('move', function (player) {
  socket.broadcast.emit('playermoved', leanify(player));
})
```

A player used the chat feature

In order to bring greater interactivity between players, we will implement a chat feature in our game. On the server side, very little code is required. This is much similar to the previous player action, where we simply broadcast the playersaid event to all the other players:

```
socket.on('speak', function (player) {
  socket.broadcast.emit('playersaid', leanify(player));
});
```

No extra information is required since the player object will contain all of the information the other clients need. Again, this is a direct benefit of using JavaScript on the server side. The server can simply receive the objects and pass them back to other clients without having to do any processing on them.

Sparing network traffic

While the Node.js and Titanium combination makes it easier to pass complex objects from one end to the other without a single line of code, passing those big and complex objects back and forth can be costly in terms of network traffic, especially on mobiles where bandwidth can be limited and not always reliable. This has a severe impact on the overall performance resulting in a poor user experience.

As you might remember from *Chapter 6, JRPG – Second to Last Fantasy*, the players array is populated with Character objects. These objects can become pretty big since they hold references to properties that are related to the game engine and things such as SpriteSheet, animations, and so on. While all the properties make sense to the client, there is no reason for them to transit back and forth to the server.

This is why we will create a function called leanify that will isolate only the relevant information about the Character object and return a new object with a much smaller set of properties. This new "lean" object will be far less complex, but it will contain all the information required for our multiplayer game.

It is important to check for the existence of every property before assigning it to the returned object. This will avoid errors related to undefined objects:

```
function leanify(p) {
  var leanPlayer = { id: p.id };

  if (p.x) {
    leanPlayer.x = p.x;
  }

  if (p.y) {
    leanPlayer.y = p.y;
  }

  if (p.direction) {
    leanPlayer.direction = p.direction;
  }

  if (p.image) {
    leanPlayer.image = p.image;
  }

  if (p.caption) {
    leanPlayer.caption = p.caption;
  }

  return leanPlayer;
}
```

Another very important reason behind this function's existence is the multiplatform nature of our game. As we have mentioned earlier, Character objects can be linked to properties that are linked to the native game engine module. Although both the versions share the same module ID, they are fundamentally different at their core. This means that if an iOS player were to log in to another player's game that is running Android, the application is bound to crash since it will try to instantiate properties that don't even exist on its platform. Hence, there is a need for such a function that only retrieves the functional aspects of the game and not the technical inner workings.

Make sure everything runs smoothly

Now is the time to do a final test run on our game server. In the command line, provided that there is no syntax error in our code, we should see the following output:

```
$ node app.js
   info  - socket.io started
Listening at: http://localhost:8080
```

We now have a fully-functional game server ready to host online multiplayer games.

Let's bring this game online!

At the beginning of this chapter, we copied assets from the previous chapter. This game is fully-functional, yet it is a single player game only. In the following section, we will expand on this game in order to make it a multiplayer game:

- We need to communicate with the game server we just created in the previous section
- We also want players to be able to exchange short messages through a rudimentary chat
- And since there will be many players roaming on the map at the same time, we will allow players to change their hero's appearance in order to have a clear distinction between players

Now, there will be a lot of changes to apply throughout the code base, but we will go through them step-by-step.

Connecting to the server

Since this is a multiplayer game, the first thing we will do is create a connection to our newly created game server.

To do this, we will load the `socket.io` module located in our `Resources` directory. Just as we load any other CommonJS file, we will use the `require` function and store the reference to the module in a variable named `io` as follows:

```
var io = require('socket.io'),
```

Then, we will create the connection with the server itself using the `connect` function. This will return a web socket object that we will be using later to interact with the server. Therefore, we will store its reference in a variable named `socket` as follows:

```
var socket = io.connect('ws://yourgameserver.com:8080/');
```

> Keep in mind that when working with your development server, you cannot connect to your game server using localhost as the server host (you cannot use 127.0.0.1 either). The reason behind this is quite understandable actually. If you are testing your app from the simulator/emulator, localhost will point to the virtual device's IP address. Even though it is running on your desktop, it will still "believe" it has its own IP address and hostname. The solution for that is to simply enter the IP address of the actual machine running the server.

Every player is different

Since we want to give players the ability to select their hero's appearance, we will perform the following steps:

1. Create new assets for these new appearances.
2. Modify the `Character` class to make the appearance dynamic.
3. Provide a way for our players to select the desired appearance.
4. Be sure to relay the hero's appearance to the game server.

We will now go over each of these steps.

Designing the hero selection window

Since selecting the hero must be the first thing that a player should do when joining a game, we will present a view that will show all the appearances available when the application starts. This view will act as a pop up and will be comprised of a label for the title and a ScrollableView component to display the previews of the hero. A ScrollableView component encapsulates a given set of child views. Each view acts as a page and only one page is shown at any given time. Users can navigate from one page to another using horizontal swipe gestures. Finally, we will add a single button at the bottom to validate the player's choice. Note that the view will not occupy the whole height of the screen; in order to give a modal pop-up effect.

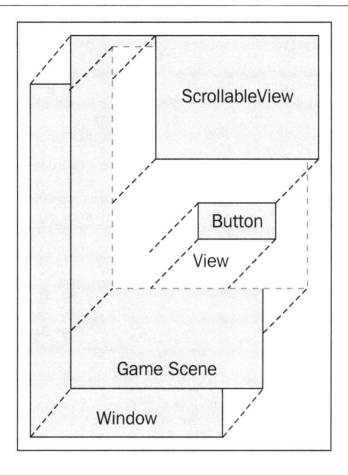

 One question that comes to mind is: "Why a view and not a window?" While both can achieve the same goal, the view is mostly chosen for aesthetic reasons, and also for simplicity, because showing a new window right before the main window is displayed would require us to change the application's navigation. Finally, creating customized dialog windows is allowed on Android, but that is not the case with iOS. This allows us to have a cross-platform look and feel.

Gathering new assets

We now need to gather the necessary assets for our hero's other appearances. To maintain the theme, we will also use SpriteSheets from Sith Jester's website (you can look up this website in the *Appendix , References*). From there, we will choose 10 more new appearances for our hero and save their respective SpriteSheets into our `assets` directory.

We will also retrieve a preview of every hero's image so that we can show it in the hero selection window. The previewed images share the same name as their SpriteSheets' counterparts on the website (only in different directories). We will rename our preview files using the `_preview` suffix at the end, for example, the `knight_m.png` SpriteSheet will have a preview image named `knight_m_preview.png`. This will give our players the ability to select one among the following 11 appearances for their hero, giving each player a distinct identity online:

To the code!

Since we want to isolate all the behavior from our hero selection windows into a separate file, we will create a new file named `hero_select.js` in our `Resources` directory.

The first thing we will do is define all the names of the previewed images that we will display in the scrollable view and store them into an array variable named `heroes` as follows:

```
var heroes = [
  'archer_m_preview.png',
  'dancer_preview.png',
  'dragoon_m_preview.png',
  'knight_f_preview.png',
  'knight_m_preview.png',
  'mediator_f_preview.png',
  'pirate_f_preview.png',
  'redmage_m_preview.png',
  'squire_f_preview.png',
  'squire_m_preview.png',
  'summoner_f_preview.png'
];
```

If we ever want to add more appearances for our hero, all we need to do is copy the image files into the `assets` directory and simply add an element to the array so that the game can display it from the list. Another approach to populate the `heroes` array would be listing all the preview files from the `assets` directory.

We will then create a function called `HeroSelectionView` that will return our encapsulated view with all the associated controls. This function will take a single parameter, which will be a `callback` function. This `callback` function will be called once the player has made his selection. This is pretty much similar to the event listeners we have covered throughout the book, where you pass a function that will be automatically called when certain conditions are met:

```
function HeroSelectionView(callback) {
```

We will then create our view that will hold all the UI elements from the dialog. It will have a `black` background, but will have a slight translucent effect (`90%` to be exact). Next, we want our view to occupy three quarters of the screen's height and we want it to be above any graphical element displayed in the game scene (hence we have given a high `zIndex` value). Since we will be adding other UI components into this view, we will store its reference into the `self` variable for later use as follows:

```
var self = Ti.UI.createView({
  backgroundColor: 'black',
  opacity: 0.9,
  height: '75%',
  zIndex: 100
});
```

We will then create a label that will act as our pop-up window's title. It will be placed at the top of the view. It will span the whole width of its parent view and the height will adapt depending on the label's content. We will give it a "purple-like" background color as well as a big font so that it stands out. Finally, we will add this label to our parent view as follows:

```
var title = Ti.UI.createLabel({
  text: 'Choose your hero',
  top: 0,
  width: '100%',
  height: Ti.UI.SIZE,
  backgroundColor: '#4B0082',
  color: 'white',
  font: {
```

```
        fontSize: '25sp',
        fontWeight: 'bold'
    }
});

self.add(title);
```

As we mentioned earlier, a scrollable view component scrolls through other views by horizontal swipe gestures. To do this, we will declare an array that will contain all of our created image views, named `heroImages`. We will then loop through our `heroes` array and create one image view per preview image. Once created, we will simply add the newly created image view to our `heroImages` array as follows:

```
var heroImages = [];

for (var i in heroes) {
    var img = Ti.UI.createImageView({
        image: 'assets/' + heroes[i],
        height: 250,
        width: 168
    });

    heroImages.push(img);
}
```

Now that all the image views have been created, we will proceed and create our scrollable view. We will assign the `views` property to our array containing all the image views we created earlier. It will span 70 percent of the parent view's height and will show the paging control. The paging control's background color will match the label's background for visual consistency and will have a translucent effect (much similar to the parent view).

 The scrollable view supports an on-screen paging control. On iOS, it appears as small dots at the bottom of the screen, while on Android, arrows are displayed on each side. This is very useful since it shows the user whether a previous or next page exists.

Finally, we will add the ScrollableView component to our parent view as follows:

```
scrollableHeroes = Ti.UI.createScrollableView({
  views: heroImages,
  top: 40,
  height: '70%',
  showPagingControl: true,
  pagingControlColor: '#4B0082',
  pagingControlAlpha: 0.8
});

self.add(scrollableHeroes);
```

The last control in our pop-up window is a button, which will be placed at the bottom of the view, without forgetting to add it to the parent view:

```
var btn = Ti.UI.createButton({
  title: 'Start Game',
  bottom: 20
});

self.add(btn);
```

When the player clicks on the **Start Game** button, we will retrieve the currently selected appearance of the hero (the current page), and store its value in the heroPreview variable. From there, we will call the callback function (we will be using it later) by passing the SpriteSheet selected by the player. To do this, we will simply remove the _preview suffix from the filename. Finally, we will hide the view in order to return to the game as follows:

```
btn.addEventListener('click', function(e) {
  var heroPreview = heroes[scrollableHeroes.currentPage];

  callback(heroPreview.replace('_preview', ''));
  self.hide();
});
```

The function will return the encapsulated view with all its child components as follows:

```
  return self;
}
```

As we did with every CommonJS module so far, we will export the function so that it can be accessed from other components using the require function:

```
module.exports = HeroSelectionView;
```

Changing the hero's appearance

Since we now have the ability to choose a different appearance for our hero, we must be able to change the appearance. To achieve this, we need to do a slight modification to the `character.js` file. We will add two parameters to the `Character` function, that are the game scene object and the SpriteSheet respectively. We will then use this second parameter in the function while creating our sprite sheet. We will also add the SpriteSheet to the game scene, as given in the following code:

```
function Character(scene, spriteSheet) {

  var self = quicktigame2d.createSpriteSheet({
    image: 'assets/' + spriteSheet,
    width: 32,
    height: 48,
    border: 0,
    margin: 1
  });
    scene.add(self);
    ...
```

The hero has no clothes

We will now update our existing code to reflect our changes to the `Character` object. Remember, the constructor now expects two parameters, and although this can be checked at many levels, we will simply update only one instance where it is referenced. So, in the `app.js` file, we will find and update the following statement by adding the two parameters (the game scene object and the default appearance):

```
var hero = new Character(scene, 'knight_m.png');
```

Since the `Character` class automatically adds itself to the game `scene` object (using the parameter), there is no need to add the hero to our `scene` object manually. So, we will remove the following line from our `app.js` file:

```
scene.add(hero);
```

Making this work

Now that everything is in place, we will call our hero selection window, which is the first thing we need to do when the game starts. To do this, we will update the `app.js` file and then move on to the `onload` event listener:

```
game.addEventListener('onload', function(e) {
  ...
  game.start();
```

We want to add the view to the right after the game starts. So, we will insert the following code right on the next line.

We will load the `HeroSelectionView` module using the `require` function and store its reference in a variable with the same name as follows:

```
var HeroSelectionView = require('hero_select');
```

We will now create a function that will be called when the player selects his hero's appearance. We will keep a reference to this callback function in a variable named `heroSelectedCallback`. In this same function, we will remove the hero that is currently present in the game scene:

```
var heroSelectedCallback = function(imageSheet) {
   scene.remove(hero);
```

We will then create a new `Character` object and assign its reference to the `hero` global variable. We will then center our hero on the screen and set its z coordinate so that it remains at the same level as all the other characters on the screen:

```
hero = new Character(scene, imageSheet);
centerHero();
hero.z  = 2;
```

Once our newly dressed hero is created, we will join the online game by assigning a unique identifier to our hero. To do this, we will use the `Titanium.Platform.id` property that represents the application's globally unique ID (UUID), ensuring that every player that connects to the server is truly unique:

```
hero.id = Titanium.Platform.id;
```

The very last thing that we will do in this callback function is actually join the game. We will do this by emitting a join event and passing our hero's unique ID as well as his appearance (note that we stripped the `assets` directory since we won't need it):

```
socket.emit('join', {
   id: hero.id,
   image: hero.image.replace('assets/', '')
});
}
```

Now that our callback function is properly defined, we will simply create the `HeroSelectionView` object and add it to our main window as follows:

```
var heroView = new HeroSelectionView(heroSelectedCallback);

win.add(heroView);
});
```

As the official Titanium documentation states, the application's unique ID can be many different things depending on the platform. On Android, this may be the **UDID (unique device ID)**. For iOS, it is a unique identifier for the installation of this application. While there is little to no certainty of knowing how exactly those unique identifiers are generated, just keep in mind that they are the most reliable way to make sure every installation is unique.

Before we go any further

Since we created a whole new JavaScript file and edited many parts of the existing code, it would be wise to make sure everything works and that (hopefully) we haven't broken anything. Once again, we will click on the **Run** button from the **App Explorer** tab. From there, we should be presented with our hero selection window right when the game starts. Once the appearance is selected and the button has been clicked, we should see our new hero in the middle of the screen harboring his new appearance, as shown in the following screenshot:

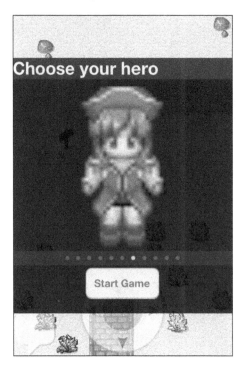

Making our hero speak

In order to bring more interactivity to our game, we want to let our players exchange short phrases. This feature is pretty straightforward. If a player wants his hero to speak, he will simply click on a button. From there, a small dialog window will ask the player to enter the desired text. Once the text is submitted, it will appear over the hero's head (similar to most online games).

Back to the drawing board

Even though the interface for the dialog window is quite simple, we will go over it briefly so that we have a clear idea of its structure. Much like the hero selection window, it will appear as an overlay of the ongoing game. It comprises of a label, a **TextArea**, and a **Button** at the bottom, as shown in the following diagram:

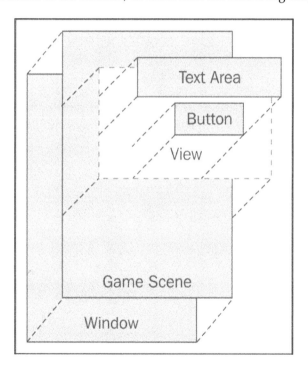

Again, we want to have a dedicated file for out chat window. Therefore, we will create a new file called `chat.js` in our `Resources` directory. In this same file, we will declare a `ChatView` function that will take a `callback` function as the parameter. Again, this `callback` function will be called when the user submits his message:

```
function ChatView(callback) {
```

We will create a view with a dark background that has a slight translucent effect. It will be `170` dp high and we will give its z coordinate a value of `100`, so that nothing overlaps the dialog window. Since we don't want to display our view right away, we will make it invisible and toggle the visibility flag as needed:

```
var self = Ti.UI.createView({
  top: 150,
  backgroundColor: 'black',
  opacity: 0.9,
  height: 170,
  zIndex: 100,
  visible: false
});
```

We will then create a label that acts as the title bar for our dialog window with a white, bold font and purple background. We will then add it to our parent view as follows:

```
var title = Ti.UI.createLabel({
  text: 'Your hero will speak!',
  top: 0,
  width: '100%',
  height: Ti.UI.SIZE,
  backgroundColor: '#4B0082',
  color: 'white',
  font: {
    fontSize: '25sp',
    fontWeight: 'bold'
  }
});

self.add(title);
```

We then need to create a text area so that players can enter the message text. A text area is nothing more than a multiline text field, and since the input text can be longer, it makes perfect sense to use such a component. It will span 90% of the parent view's width and be 75 dp high. It will have a thick, white border and rounded corners, thanks to the `borderRadius` property. It will have a big, white letter font with a black background and messages will be limited to 25 characters. We will then add it to the parent view as follows:

```
var txtSay = Ti.UI.createTextArea({
  top: 37,
  width: '90%',
  height: 75,
  borderWidth: 3,
  borderRadius: 4,
  borderColor: '#fff',
  color: '#fff',
  backgroundColor: '#000',
  maxLength: 25,
  font: {
    fontSize: '22sp'
  }
});

self.add(txtSay);
```

The last UI element of the view will be a simple button that will be placed at the bottom of the view. Again, we will add it to the parent view as follows:

```
var btnSay = Ti.UI.createButton({
  title: 'Say It!',
  bottom: 7
});

self.add(btnSay);
```

When the player clicks on the **Say It!** button, we will first call the `blur` function on the text area. This will hide the onscreen keyboard. Then, we will invoke the `callback` function by passing it the input field's value as follows:

```
btnSay.addEventListener('click', function(e) {
  txtSay.blur();
  callback(txtSay.value);
```

We will then reset the text area's value so that it is not present when the view is displayed again later in the game. Finally, we will hide the view so that the game can move on:

```
txtSay.value = '';

self.hide();
    });
```

The `ChatView` function will return the `self` parent view since it contains all of the UI components:

```
return self;
}
```

Since we want to invoke this function using the CommonJS pattern, we will export the function from the module as follows:

```
module.exports = ChatView;
```

How to reach the window

We might have a working chat window, but right now, there is no way to invoke it from the game. Let's remedy by creating a button to access it. But at this point, keep in mind that we are in the middle of the game and that we don't want to use native controls inside a game scene.

Therefore, in our `app.js` file, we will create a new sprite with a little bubble image. It will be translucent just like the V-Pad and it will have the z coordinate as `5`:

```
var chat_button = quicktigame2d.createSprite({
    image:'assets/chat.png'
});
chat_button.alpha = 0.5;
chat_button.z = 5;
```

We also need to add it to our game scene object in order for it to be displayed on the screen:

```
scene.add(chat_button);
```

You may have noticed that we did not specify any position on the screen for our new sprite. That is because the screen dimensions are not set until the `onload` event occurs. So, we will go into the event handler and set our `chat_button` position to the bottom left-hand corner of the screen:

```
game.addEventListener('onload', function(e) {
    ...
    chat_button.y = game.screen.height - chat_button.height;
```

Wiring it up

With our **Chat** button on the screen, we can now detect when players touch it by showing our chat window. We will load the `ChatView` module using the require function:

```
var ChatView = require('chat');
```

We will then create a new instance of the `ChatView` class by passing a very basic function that retrieves the message caption entered and instructs the hero with this same message. Also, it informs the server that your hero said something in order to notify the other players. Since our chat window is basically a view, it must be added to the main application window:

```
var chatView = new ChatView(function(caption) {
  hero.say(caption);

  socket.emit('speak', {
    id: hero.id,
    caption: caption
  });
});


win.add(chatView);
```

Since our **Chat** button and the V-Pad share the same touch principles, both can share the same event listener and take different actions depending on which sprite is touched. What we basically do here is check whether the touch coordinates are contained within the bounds of the **Chat** button. If that is the case, we will then show the ChatView instance:

```
game.addEventListener('touchstart', function(e) {
  ...
  if (chat_button.contains(touchX, touchY)) {
    chatView.show();
  }
```

We will now take our latest code for a spin using the **Run** button from the **App Explorer** tab and we should see our new **Chat** Button at the bottom-left corner of the screen. When we tap it, the chat window will be displayed allowing us to enter our message, as shown in the following screenshot:

Displaying what the hero is saying

The chat functionality is still incomplete because there is no way to actually see what our hero is saying. The reason for that is pretty obvious. In the `click` event handler of the **Say It!** button, we call the `say` function from the `Character` class. And that function is nowhere to be found in the actual codebase. So, let's implement it right away.

In the `character.js` file, we will create a `TextSprite` object. Think of it as a label inside a game scene. We will give it a default text since the sprite's dimensions are calculated at the time of creation. It will be centered and have the z coordinate as 2:

```
var textsprite = quicktigame2d.createTextSprite({
  text: 'Lorem ipsum dolor sit amet',
  fontFamily: 'Verdana',
  fontSize: 14,
  textAlign: Ti.UI.TEXT_ALIGNMENT_CENTER,
  z: 2
});
```

Since we want to avoid confusion in determining who is talking, we will set a random color for every new character on screen. We also need to hide the sprite until the player decides to say something. Just as we did for the SpriteSheet, we will add this new sprite to the game scene as follows:

```
textsprite.color(Math.random(), 0, 0);
textsprite.hide();
scene.add(textsprite);
```

We can now add the say function to our class. The basics of this function are pretty simple. We will update the sprite's text with the new caption. We will position the sprite right above our hero's head and center it. We will then show the message, but since we don't want it to stay on the screen indefinitely, we will hide it after 5 seconds:

```
self.say = function(caption) {
    textsprite.text = caption;

    textsprite.x = self.x -(textsprite.width * 0.5);
    textsprite.y = self.y - textsprite.height - 15;

    textsprite.show();

    setTimeout(function() { textsprite.hide() }, 5000);
}
```

Let us run our code one more time and see if our hero can now speak without a hitch.

Our text is there! The chat feature is now fully implemented , as shown in the preceding screenshot.

Interacting with the game server

We have now covered all of the features required to start interacting with our game server, which in turn allow us to interact with other players as well. Just as we did on the server side, our game must keep tabs on the other players in terms of who gets in, who gets out, what they are saying, and most importantly, where they are on the map.

We will keep all the connected players in an array called `players`:

```
var players = [];
```

All server interactions use the player's unique identifier to determine who does what. We will also create a utility function named `getPlayer` that will retrieve the player matching a given ID from the array:

```
function getPlayer(id) {
  for (p in players) {
    if (players[p].id === id) {
      return players[p];
    }
  }
}
```

When our hero moves

We need to inform the server when our hero changes his position around the map. To do this, all we have to do is to send the new x and y coordinates to the server when the hero moves. This is done in the `updateVpad` function. Right at the end of the movement block, we will emit a `move` event to the server by passing the hero's unique ID, the absolute x and y positions (the overall position on the map), as well as the heading direction. This way, the other players will be able to see the direction in which we are heading:

```
function updateVpad() {
  if (isVpadActive) {
    ...
    socket.emit('move', {
      id: hero.id,
      x: Math.round(Math.abs(hero.x + map.x)),
      y: Math.round(Math.abs(hero.y + map.y)),
      direction: heroDirection
    });
  } else {
    ...
  }
}
```

Notice that we rounded the numbers for the x and y values. The reason is that the `quicktigame2d` module uses floating values for its coordinates, which amounts to a lot of decimals for greater precision. Since we don't really need that kind of precision in our game, rounding it reduces the amount of information transited without having any impact on the game.

This looks familiar

If you look more closely at the game server code, you will find that it emits four different types of events, which are as follows:

- `playerjoined`
- `playerquit`
- `playermoved`
- `playersaid`

These four events (although named differently), cover the same four types of events our game sends when the player takes action. All we have to do now, is implement the proper handlers and we will be able to process the action taken by the other players.

Someone joined

When a new player joins the game, we will create a new `Character` object using the image (appearance) this new player will have. In order to be able to recognize it down the line, we will assign it a unique ID, which is provided by the `player` variable passed by the server. After setting its z coordinate, we will push this newly created character into our `players` array:

```
socket.on('playerjoined', function(player) {
  var newPlayer = new Character(scene, player.image);

  newPlayer.id = player.id;
  newPlayer.z = 2;

  players.push(newPlayer);
});
```

Someone quit

When a player quits the game, we will loop through all the connected players until we find the one that matches the unique ID passed from the server. Once found, we will make the leaving player say **Goodbye!** and then remove the player from the game scene. Finally, we will remove the player from the `players` array:

```
socket.on('playerquit', function(player) {
  for (p in players) {
    if (players[p].id === player.id) {
      players[p].say('Goodbye!');
      scene.remove(players[p]);
      players.splice(p);
    }
  }
});
```

Note that we could not use the `getPlayer` function to retrieve the element to be removed, because JavaScript's `splice` function expects the position and not the object.

Someone moved

When a player moves on the map, the first thing we will do is retrieve the player that moved locally using the `getPlayer` function. If a player is found, we will update the absolute position as well as the direction based on the information provided by the server:

```
socket.on('playermoved', function(player) {
  var p = getPlayer(player.id);

  if (p) {
    p.absolute_x = player.x;
    p.absolute_y = player.y;
    p.direction = player.direction;
```

Once the absolute position is populated, we can now calculate the relative position on the screen, since the map may be offset due to our hero's movements:

```
    p.x = p.absolute_x + Math.round(map.x);
    p.y = p.absolute_y + Math.round(map.y);
  }
});
```

Someone spoke

When another player says something, all that we will do is retrieve the player matching the unique ID, and call the `say` function with the incoming message caption as the parameter:

```
socket.on('playersaid', function(player) {
  var p = getPlayer(player.id);

  p.say(player.caption);
});
```

Where is everybody?

If we want to run our game right now, it would work flawlessly; if we want to look at the server's console, we would see that the information flows between the server and the connected client. There is only one caveat; we don't see any other player on the screen besides our hero.

The reason is simple. Even though we assigned the other players' coordinates when we received the events from the server, we didn't position them on the screen as our hero was roaming the map. To achieve this, we will create a `drawOtherPlayers` function that will loop through all of the connected players and will update their x and y positions (even if the other players don't move, our hero will move and change the map's offset at the same time. This is why we must redo these calculations):

```
function drawOtherPlayers() {
  for (p in players) {
    var other = players[p];

    other.x = other.absolute_x + Math.round(map.x);
    other.y = other.absolute_y + Math.round(map.y);
```

Now, the following condition might seem complex, but in fact, its purpose is actually quite simple. It is used to determine if any other player is currently present in the Viewport. If that is the case, we will update his direction so that he or she faces the direction passed by the server. If he is not in the Viewport, there is no need to update the direction. This will save the processing time to achieve the operation:

```
if (other.id !== Ti.Platform.id) {
  if (((other.x + other.width/2) >= 0)
    && ((other.x - other.width/2) <=
      game.screen.width)
    && ((other.y + other.height/2) >= 0)
    && ((other.y - other.height/2) <=
    game.screen.height)) {

    other.halt();
    }
  }
 }
}
```

> In game development, it is considered good practice to only do operations on elements that are visible on the screen. Or, at least do the bare minimum to keep tabs on objects that are not visible. This greatly reduces the overhead and improves performance and can even become a necessity while dealing with resource-intensive games.

We will now invoke our new function inside the `updateVpad` function:

```
function updateVpad() {
  if (isVpadActive) {
    ...
    drawOtherPlayers();
```

Be sure to hang up that connection

Once we finish playing our game, we need to inform the server that we are quitting the game. To do this, we will create an event listener on the `close` event on the main window. Inside this handler, we will simply emit a `quit` event with our `hero` as the parameter and then let the game server do the rest:

```
win.addEventListener('close', function() {
  socket.emit('quit', hero);
});
```

Game on!

We can now move on to test the entire solution using different clients and devices test all the different interactions. One of the easiest ways of testing more than one client on a single machine is to run the iPhone simulator and the Android emulator concurrently. Of course, the testing on device still remains the best way to test such an application.

Summary

This was the broadest and longest chapter in the book, and we covered a lot of concepts. We first learned about Web Sockets and how they can be used in a mobile application. We also learned about Node.js, its installation, how it works, and its package management system (npm). We then went on to code a complete game server using JavaScript by leveraging the `socket.io` library.

We integrated another native extension module into our application, even though it does not share the same version number between platforms. We also added a large number of graphical assets into our game and managed to use them in a dynamic manner.

Finally, we went over how to interact with our game server from our mobile application and keep tabs on the other player's actions.

Once all that has been written and done, we now have a working online multiplayer game, and what was learned in this chapter can easily be applied to many other types of online games.

In the next chapter, we will learn how to make our applications interact with popular social networks such as Facebook and Twitter.

8
Social Networks

Social networks have gone from a novelty to something that almost everyone uses on a daily basis. There are many networks out there targeted at different audiences, and it is not uncommon that someone uses more than one network. This chapter will address this issue by providing our user with a way to post the same status message on both Facebook and Twitter, with the click of a single button.

By the end of this chapter, you will have learned the following concepts:

- Creating and configuring applications using developer accounts
- Handling device rotation
- Interacting with Facebook using Titanium's provided module
- Interacting with Twitter using a CommonJS library
- Creating the **Settings** window using configuration files to adhere to platform standards
- Using native Android menus

Creating our project

As we did in our previous projects, we need to set up a new project for our application. To do this, navigate to **File | New | Mobile Project** from Titanium Studio and fill out the wizard forms with the following information:

Field	Value to be entered
Project Template	Classic, the default project
Project Name	`Unified Status`
Location	You can either: • Create the project in your current workspace directory by checking the **Use Default Location** checkbox • Create the project in a location of your choice
App Id	`com.packtpub.hotshot.unifiedstatus`
Company/Personal URL	`http://www.packtpub.com`
Titanium SDK Version	By default, the wizard will select the latest version of the Titanium SDK as this is recommended (while writing this book, we were using Version 3.1.3 GA)
Deployment Targets	Check **iPhone** and **Android**
Cloud Settings	Uncheck the **Cloud-enable this application** checkbox

Project creation is covered in more extensive detail in *Chapter 1, Stopwatch (with Lap Counter)*. So feel free to refer to this section if you want more information regarding project creation.

One window to rule them all

Since we want to give our users the ability to post their messages on multiple social networks at once, it makes perfect sense to keep our whole application in a single window. It will be comprised of the following sections:

- The top section of the window will contain labels and a text area for message input. The text area will be limited to 140 characters in order to comply with Twitter's message limitation. As the user types his or her message, a label showing the number of characters will be updated.

- The bottom section of the window will use an encapsulating view that will contain multiple image views. Each image view will represent a social network to which the application will post the messages (in our case, Twitter and Facebook). But we can easily add more networks that will have their representation in this section. Each image view acts as a toggle button in order to select if the message will be sent to a particular social network or not.

- Finally, in the middle of those two sections, we will add a single button that will be used to publish the message from the text area.

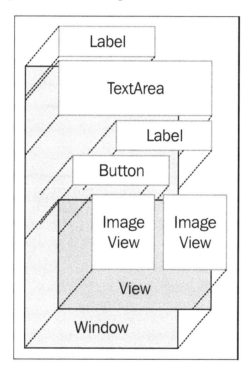

Code while it is hot!

With our design in hand, we can now move on to create our user interface. Our entire application will be contained in a single window, so this is the first thing we will create. We will give it a title as well as a linear background gradient that changes from purple at the top to black at the bottom of the screen. Since we will add other components to it, we will keep its reference in the `win` variable:

```
var win = Ti.UI.createWindow({
  title: 'Unified Status',
  backgroundGradient: {
    type: 'linear',
    startPoint: { x: '0%', y: '0%' },
    endPoint: { x: '0%', y: '100%' },
    colors: [ { color: '#813eba'}, { color: '#000' } ]
  }
});
```

The top section

We will then add a new white label at the very top of the window. It will span `90%` of the window's width, have a bigger font than other labels, and will have its text aligned to the left. Since this label will never be accessed later on, we will invoke our `createLabel` function right into the `add` function of the window object:

```
win.add(Ti.UI.createLabel({
  text: 'Post a message',
  color: '#fff',
  top: 4,
  width: '90%',
  textAlign: Ti.UI.TEXT_ALIGNMENT_LEFT,
  font: {
    fontSize: '22sp'
  }
}));
```

We will now move on to create our text area where our users will enter their messages. It will be placed right under the label previously created, occupy 90% of the screen's width, have a slightly bigger font, and have a thick, rounded, dark border. It will also be limited to 140 characters. It is also important that we don't forget to add this newly created object to our main window:

```
var txtStatus = Ti.UI.createTextArea({
  top: 37,
  width: '90%',
  height: 100,
  color: '#000',
  maxLength: 140,
  borderWidth: 3,
  borderRadius: 4,
  borderColor: '#401b60',
  font: {
    fontSize: '16sp'
  }
});

win.add(txtStatus);
```

The second label that will complement our text area will be used to indicate how many characters are currently present in the user's message. So, we will now create a label just underneath the text area and assign it a default text value. It will span 90% of the screen's width and have its text aligned to the right. Since this label will be updated dynamically when the text area's value changes, we will keep its reference in a variable named lblCount. As we did with our previous UI components, we will add our label to our main window using the following code:

```
var lblCount = Ti.UI.createLabel({
  text: '0/140',
  top: 134,
  width: '90%',
  color: '#fff',
  textAlign: Ti.UI.TEXT_ALIGNMENT_RIGHT
});

win.add(lblCount);
```

The last control from the top section will be our **Post** button. It will be placed right under the text area and centered horizontally using the following code:

```
var btnPost = Ti.UI.createButton({
  title: 'Post',
  top: 140,
  width: 150
});


win.add(btnPost);
```

 By not specifying any left or right property, the component is automatically centered. This is pretty useful to keep in mind while designing our user interfaces, as it frees us from having to do calculations in order to center something on the screen.

Staying within the limits

Even though the text area's maxLength property will ensure that the message length will not exceed the limitation we have set, we need to give our users feedback as they are typing their message. To achieve this, we will add an event listener on the change event of our text area. Every time; the text area's content changes, we will update the lblCount label with the number of characters our message contains:

```
txtStatus.addEventListener('change', function(e) {
  lblCount.text = e.value.length + '/140';
```

We will also add a condition to check if our user's message is close to reaching its limit. If that is the case, we will change the label's color to red, if not, it will return to its original color:

```
if (e.value.length > 120) {
  lblCount.color = 'red';
} else {
  lblCount.color = 'white'
}
```

Our last conditional check will be enabling the **Post** button only if there is actually a message to be posted. This will prevent our users from posting empty messages:

```
  btnPost.enabled = !(e.value.length === 0);
});
```

Setting up our Post button

Probably the most essential component in our application would be the **Post** button. We won't be able to post anything online (yet) by clicking on it, as there are things we will still need to add.

We will add an event listener for the `click` event on the **Post** button. If the on-screen keyboard is displayed, we will call the `blur` function of the text area in order to hide it:

```
btnPost.addEventListener('click', function() {
    txtStatus.blur();
```

Also, we will reset the value to the text area and the character count on the label, so that the interface is ready to enter a new message:

```
    txtStatus.value = '';
    lblCount.text = '0/140';
});
```

The bottom section

With all our message input mechanisms in place, we will now move on to our bottom section. All components from this section will be contained in a view that will be placed at the bottom of the screen. It will span 90% of the screen's width, and its height will adapt to its content. We will store its reference in a variable named `bottomView` for later use:

```
var bottomView = Ti.UI.createView({
    bottom: 4,
    width: '90%',
    height: Ti.UI.SIZE
});
```

We will then create our toggle switches for each social network that our application interacts with. Since we want something sexier than regular switch components, we will create our own switch using a regular image view.

Our first image view will be used to toggle the use of Facebook. It will have a dark blue background (similar to Facebook's logo) and will have a background image representing Facebook's logo with a gray background, so that it appears disabled. It will be positioned to the left of its parent view, and will have a `borderRadius` value of 4 in order to give it a rounded aspect. We will keep its reference in the `fbView` variable and then add this same image view to our bottom view using the following code:

```
var fbView = Ti.UI.createImageView({
  backgroundColor: '#3B5998',
  image: 'images/fb-logo-disabled.png',
  borderRadius: 4,
  width: 100,
  left: 10,
  height: 100
});


bottomView.add(fbView);
```

We will create a similar image view (in almost every way), but this time for Twitter. So the background color and the image will be different. Also, it will be positioned to the right of our container view. We will store its reference in the `twitView` variable and then add it to our bottom view using the following code:

```
var twitView = Ti.UI.createImageView({
  backgroundColor: '#9AE4E8',
  image: 'images/twitter-logo-disabled.png',
  borderRadius: 4,
  width: 100,
  right: 10,
  height: 100
});


bottomView.add(twitView);
```

Last but not the least, it is imperative that we do not forget to add our `bottomView` container object into our window. Also, we will open the window so that our users can interact with it using the following code:

```
win.add(bottomView);
win.open();
```

What if the user rotates the device?

At this stage, if our user were to rotate the device (to landscape), nothing would happen on the screen. The reason behind this is because we have not taken any action to make our application compatible with the landscape mode. In many cases, this would require some changes to how the user interface is created depending on the orientation. But since our application is fairly simple, and most of our layout relies on percentages, we can activate the landscape mode without any modification to our code.

To activate the landscape mode, we will update the `orientations` section from our `tiapp.xml` configuration file. It is mandatory to have at least one orientation present in this section (it doesn't matter which one it is). We want our users to be able to use the application, no matter how they hold their device:

```
<iphone>
  <orientations device="iphone">
    <orientation>Ti.UI.PORTRAIT</orientation>
    <orientation>Ti.UI.UPSIDE_PORTRAIT</orientation>
    <orientation>Ti.UI.LANDSCAPE_LEFT</orientation>
    <orientation>Ti.UI.LANDSCAPE_RIGHT</orientation>
  </orientations>
</iphone>
```

 By default, no changes are required for Android applications since the default behavior supports orientation changes. There are, of course, ways to limit orientation in the manifest section, but this subject falls out of this chapter's scope.

See it in action

We have now implemented all the basic user interface. We can now test it and see if it behaves as we anticipated. We will click on the **Run** button from the **App Explorer** tab, as we did many times before.

We now have our text area at the top with our two big images views at the bottom, each with the social network's logo. We can already test the message entry (with the character counter incrementing) by clicking on the **Post** button.

Now if we rotate the iOS simulator (or the Android emulator), we can see that our layout adapts well to the landscape mode.

To rotate the iOS simulator, you need to use the **Hardware** menu or you can use *Cmd-Left* and *Cmd-Right* on the keyboard. If you are using the Android emulator, there is no menu, but you can change the orientation using *Ctrl + F12*.

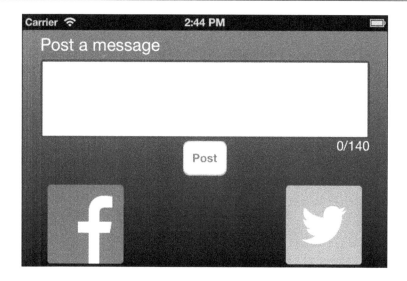

> The reason this all fits so well is because most of our dimensions are done using percentages. That means that our components will adapt and use the available space on the screen. Also, we positioned the bottom view using the `bottom` property; which meant that it will stick to the bottom of the screen, no matter how tall it is.

Polishing it up a little

If you played a little with the application in its present state, you may have noticed that once it is displayed, the keyboard tends to stay on the screen unless you click on the **Post** button. This can become problematic in cases where the user doesn't want to post any message.

To address that, we will simply create an event listener on the `click` event for the whole window. This will allow our users to hide the on-screen keyboard only by taping anywhere on the window:

```
win.addEventListener('click', function() {
  txtStatus.blur();
});
```

Facebook Integration

Facebook today needs no introduction, as it has become the most widely used social network on the planet. Facebook provides the ability for developers to develop applications that can interact with Facebook's data. Some of those applications can even reside and be hosted inside Facebook, providing an even richer experience for the user without ever leaving the social network's website.

We will interact with Facebook via their Graph API, which will then allow us to publish open graph stories (such as status messages, photos, and so on) from our application. Understanding the inner working of those APIs is out the scope of this book, but there is a lot of documentation online (`http://developers.facebook.com/docs/reference/api`) if you ever need to dive deeper into this topic.

Creating our application

Even though our application will not be running inside the Facebook website, we must declare it so that it is recognized by the API. To do this, once logged into Facebook, we will navigate to the `https://developers.facebook.com/apps` website. We will be shown a page listing all the applications we created (if any).

On this page, we will click on the **+ Create New App** button, and a form similar to the one shown in the following screenshot will be displayed:

Fill in the form using the following table to provide Facebook with the information about our app:

Field name	What Facebook expects
App Name	A unique name for your application (**Unique Status**).
App Namespace	The application namespace used for defining customized open graph actions and objects (for example, `namespace:action`) and for the URL for Apps on Facebook (for example, `http://apps.facebook.com/namespace`).
	We will not be using any specific actions involving namespaces, so we will leave this field blank.
App Category	This is the category that best describes our application. Facebook provides category guidelines to help developers determine where their application should be in case of uncertainty (**Lifestyle**).
Web Hosting	This field should remain unchecked since we have no interest in hosting any web code for this application.

Retrieving our Facebook app ID

Now that our application is created on Facebook's servers, we need to retrieve the technical application identifier (or app ID, in short). It is basically a string that will be used by our mobile application while communicating with Facebook.

It is usually displayed right under the application's name on the detail's page, as shown in the following screenshot:

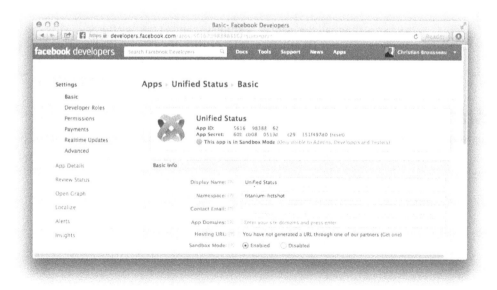

This is the only operation we will be performing on the Facebook developer's website. We have now met all the requirements for interacting with Facebook from our application.

There is a module for that

Since we don't want to interact with the social network through manual asynchronous HTTP requests, we will leverage the Facebook native module provided with Titanium. Since it comes with Titanium SDK, there is no need to download or copy any file.

All we need to do is add a reference (one for each target platform), in the `modules` section of our `tiapp.xml` file as follows:

```
<modules>
  <module platform="android">facebook</module>
  <module platform="iphone">facebook</module>
</modules>
```

Also, we need to add the following property at the end of our configuration file, and replace the FACEBOOK_APPID parameter with the application ID that was provided when we created our app online.

```
<property name="ti.facebook.appid">[FACEBOOK_APPID]</property>
```

 Why do we have to reference a module even though it comes bundled with the Titanium framework? Mostly, it is to avoid framework bloat; it is safe to assume that most applications developed using Titanium won't require interaction with Facebook. This is the reason for having it in a separate module that can be loaded on demand.

Instantiating our module

Now that our native module is configured for our project, we will load it into our code. In our `app.js` file, we will load the module using the `require` function and store its reference in a variable named `fb`:

```
var fb = require('facebook');
```

Linking our mobile app to our Facebook app

With our Facebook module loaded, we will now populate the necessary properties before making any call to the network. We will need to set the `appid` property in our code; since we have already defined it as a property in our `tiapp.xml` file, we can access it through the Properties API. This is a neat way to externalize application parameters, thus preventing us to hardcode them in our JavaScript code:

```
fb.appid = Ti.App.Properties.getString('ti.facebook.appid');
```

We will also set the proper permissions that we will need while interacting with the server. (In this case, we only want to publish messages on the user's wall):

```
fb.permissions = ['publish_actions'];
```

It is important to set the `appid` and `permissions` properties before calling the `authorize` function. This makes sense since we want Facebook to authorize our application with a defined set of permissions from the get-go.

Allowing our user to log in and log out at the click of a button

We want our users to be able to connect (and disconnect) at their will from one social network, just by pressing the same view on the screen. To achieve this, we will create a function called `toggleFacebook` that will have a single parameter:

```
function toggleFacebook(isActive) {
```

If we want the function to make the service active, then we will verify if the user is already logged in to Facebook. If not, we will ask Facebook to authorize the application using the function with the same name. If the parameter indicates that we want to make the service inactive, we will log out from Facebook altogether:

```
if (isActive) {
  if (!fb.loggedin) {
    fb.authorize();
  }
} else {
  fb.logout();
}
}
```

Now, all that we need to do is create an event listener on the `click` event for our Facebook image view and simply toggle between the two states depending whether the user is logged in or not:

```
fbView.addEventListener('click', function() {
  toggleFacebook(!fb.loggedIn);
});
```

 The `authorize` function prompts the user to log in (if he or she is not already logged in), and authorize his or her application. It kind of makes sense that Facebook requires user validation before delegating the right to post something on his or her behalf.

Handling responses from Facebook

We have now completely implemented the Facebook login/logout mechanism into our application, but we still need to provide some feedback to our users. The Facebook module provides two event listeners that allow us to track when our user will have logged in or out.

In the `login` event listener, we will check if the user is logged in successfully. If he or she did, we will update the image view's `image` property with the colored logo. If there was any error during authentication, or if the operation was simply cancelled, we will show an alert, as given in the following code:

```
fb.addEventListener('login', function(e) {
  if (e.success) {
    fbView.image = 'images/fb-logo.png';
  } else if (e.error) {
    alert(e.error);
  } else if (e.cancelled) {
    alert("Canceled");
  }
});
```

In the `logout` event listener, we will update the image view's `image` property with the grayed out Facebook logo using the following code:

```
fb.addEventListener('logout', function(e) {
  fbView.image = 'images/fb-logo-disabled.png';
});
```

Authorizing our application

Before we go any further, let's test our code and see how it behaves when we touch the disabled Facebook image view for the first time. Since we are not logged in to Facebook, we are shown the login window, as shown in the following screenshot:

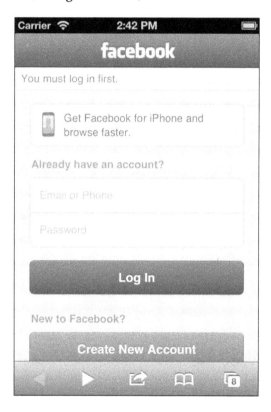

Once we have entered our credentials, we are shown a confirmation window in order to allow our application to access our public Facebook profile, and most importantly, publish messages on our behalf, as shown in the following screenshot:

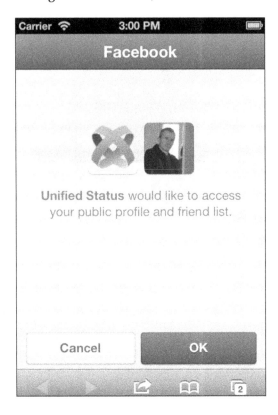

Once we have authorized the application, we should be returned to our application, and the Facebook image view should display the colored logo.

Posting our message on Facebook

Since we are now connected to Facebook, and our application is authorized to post, we can now post our messages on this particular social network. To do this, we will create a new function named postFacebookMessage, with a single parameter, and that will be the message string to be posted:

```
function postFacebookMessage(msg) {
```

Inside this function, we will call the `requestWithGraphPath` function from the Facebook native module. This function can look fairly complex at first glance, but we will go over each parameter in detail. The parameters are as follows:

- The Graph API path requested (My feed).
- A dictionary object containing all of the properties required by the call (just the message).
- The HTTP method used for this call (`POST`).
- The callback function invoked when the request completes (this function simply checks the result from the call. In case of error, an alert is displayed).

```
fb.requestWithGraphPath('me/feed', {
    message: msg
  }, "POST", function(e) {
    if (e.success) {
      Ti.API.info("Success! " + e.result);
    } else {
      if (e.error) {
        alert(e.error);
      } else {
        alert("Unknown result");
      }
    }
  }
);
}
```

We will then update the `click` event handler for the **Post** button and call the `postFacebookMessage` function if the user is logged in to Facebook:

```
btnPost.addEventListener('click', function() {
  if (fb.loggedIn) {
    postFacebookMessage(txtStatus.value);
  }
  . . .
});
```

With this, our application can post messages on our user's Facebook wall.

Twitter integration

Twitter is the most popular microblogging site in the world and has gained tremendous momentum with its short 140-character format (hence our limitation on the text area to be compatible with both sites).

Creating our application

Similar to what we did with Facebook, we need to create an application on Twitter's servers in order to be able to interact with the service. We will then navigate to their developer's website at `https://dev.twitter.com/apps`.

Once we are on this page, we will click on the **Create a new application** button. We will then be redirected to a form in order to fill in all of the information related to our application, as given in the following table:

Field name	What Twitter expects
Name	The name of our application. This name will be displayed in the authorization screen (**United Status**).
Description	A short description of our application (**Titanium Mobile Hotshot**).
Website	Although this field is required, it is only used to provide information to our users. If you don't have an official URL, you can use a generic placeholder instead.
Access	What type of access our application needs (since we want to post messages, we will select **Read and Write**).

Retrieving our consumer key and secret

Once our Twitter application is created, we need to retrieve both the **Consumer key** and **Consumer secret** strings. These values will be used by our mobile application to interact with Twitter. These two values are present in the **Details** tab of the detail page.

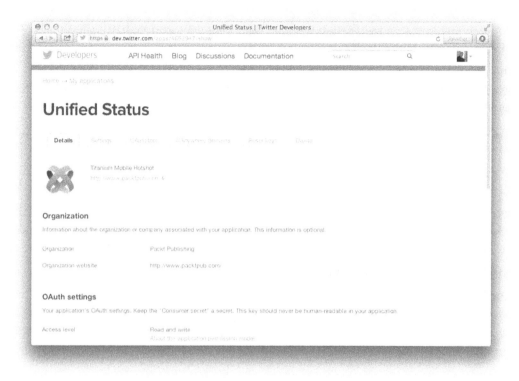

With those two values in hand, we will add them to our `tiapp.xml` configuration file. This will prevent us from hardcoding them in our JavaScript code:

```
<property name="twitter.consumerKey">
  [YOUR_TWITTER_CONSUMER_KEY]
</property>
<property name="twitter.consumerSecret">
  [YOUR_TWITTER_CONSUMER_SECRET]
</property>
```

No module, but a good, old fashioned JavaScript library

Titanium doesn't provide a native module to interact with Twitter, so we will have to rely on another mechanism. Again, since Titanium allows us to use many JavaScript libraries that respect the CommonJS pattern, we have a wide array of possibilities.

In this chapter, we will be using an open source JavaScript library called Social Plus. It was developed by a gentleman named *Aaron Saunders*, and is distributed under the Apache License Version 2.0.

The whole library is contained in a single `social_plus.js` source file and can be downloaded directly from `http://bit.ly/14AFczH`. Once downloaded, we will copy this source file in the root of our `Resources` directory.

 Twitter has made significant changes to its API in 2013. This required developers to update their codebase in order to be compatible with the latest version of the API. Therefore, it is important to be updated with the latest version while using libraries such as this one, to prevent our application from getting locked out of the API when Twitter makes changes in the future.

Instantiating the library

Since this is a JavaScript file, we invoke the library using the `require` function. We will keep its reference in a variable named `social` for future use:

```
var social = require('social_plus');
```

Linking with our Twitter application

The next step is to provide Twitter with our API credentials. To do this, we will invoke the `create` function and pass a dictionary object (containing our `consumerKey` and `consumerSecret` parameters). We will save the returned connection object into a variable aptly named `twitter`:

```
var twitter = social.create({
  consumerSecret:
    Ti.App.Properties.getString('twitter.consumerSecret'),
  consumerKey:
    Ti.App.Properties.getString('twitter.consumerKey')
});
```

Toggling the state of the Twitter connection

Now that we have a connection to the Twitter API, we will create a `toggleTwitter` function that will connect or disconnect depending on the `isActive` parameter's value. If the user wants Twitter to be active, we will call the `authorize` function from the `twitter` object (if the application is not already authorized). The `authorize` function takes a callback function as a parameter that is called when the authorization is made successfully.

If the authorization is successful, we will update the image view's `image` property with the colored logo. We will do the same if the session is already authorized:

```
function toggleTwitter(isActive) {
  if (isActive) {
    if (!twitter.isAuthorized()) {
      twitter.authorize(function() {
        twitView.image = 'images/twitter-logo.png';
      });
    } else {
      twitView.image = 'images/twitter-logo.png';
    }
```

If the user wants Twitter to be inactive, then we will call the `deauthorize` function. Also, we will update the image view's `image` property and the grayed out logo to show that it is inactive:

```
  } else {
    twitter.deauthorize();
    twitView.image = 'images/twitter-logo-disabled.png';
  }
}
```

Just as we did with the Facebook image view, we will add an event listener on the `click` event for the Twitter image view. The handler will simply toggle between the two stated views depending on whether the user is logged in or not:

```
twitView.addEventListener('click', function() {
  toggleTwitter(!twitter.isAuthorized())
});
```

Authorizing our application

With our Twitter integration pretty much functional, we can run our code and see if our mobile application can connect to Twitter successfully. Once the application is loaded, if we click on the **Twitter** Image view, we should see a confirmation window asking us to authorize the application to post messages on our behalf, as shown in the following screenshot:

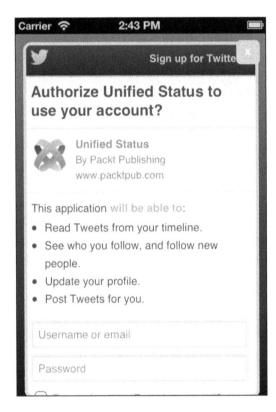

Once we authorized the application, we should return back to our application, and the Twitter image view should display the colored logo.

If the user is not logged in to the Twitter service, an authentication window will be displayed asking the user to log in to his or her Twitter account. Once logged in, the authorization window will be displayed.

Posting our message on Twitter

We will now create a new function that will post our message to our Twitter feed using the `share` function. This function is simpler than its Facebook equivalent. It takes a single dictionary object with the following three properties:

- The message content itself
- A callback function called when the message is posted successfully
- Another callback function called in case of an error while posting the message

The share function is as follows:

```
function postTwitterMessage(msg) {
  twitter.share({
    message: msg,
    success: function() {
      Ti.API.info('Tweeted!');
    },
    error: function() {
      alert('ERROR from Twitter Tweeter');
    }
  });
}
```

Since we want to be able to post to both the social networks at the touch of a button, we will update the `click` event listener for the **Post** button by adding the call to `postTwitterMessage` if the user is logged in:

```
btnPost.addEventListener('click', function() {
  ...
  if (twitter.isAuthorized()) {
    postTwitterMessage(txtStatus.value);
  }
  ...
});
```

We now have a fully, functional mobile application that allows our users to post messages on two social networks simultaneously on the click of a single button.

Settings

We will now proceed with adding a **Settings** window to our application. Most applications today have a dedicated settings section where users can modify parameters, so that the application is more adapted to their needs. Our **Settings** window will allow our users to choose their active social networks using toggle switches. We can choose between different approaches to address our requirements.

The first approach would be to develop a whole new window and manage our settings manually through code. We would have to create the user interface and manage the persistence of those settings throughout the application. Even though this might work, it would require a lot of effort and require a considerable amount of development each time we want to add a new setting. Also, iOS and Android have very different guidelines for settings management.

The second approach would be to follow on the native settings guidelines for each target platform. That would save us time since all the user interfaces and controls would be delegated to the native platform.

Before we touch anything

Since the settings are related to the `platform` directory and not to the application, all of the files will be placed in their own directory. So we will create a platform directory in the root of our project. Inside this new platform directory, we will create:

- A directory structure for Android: `/android/res/xml`
- A directory for iPhone: `/iphone`

Once created, your project's directory structure should look similar to the following screenshot:

Settings for iOS

With our directory structure in place, we will now edit our **Settings** window using native tools provided by Apple. To do this, we need to create a platform-specific file named `Settings.bundle`. While we can create such a file from scratch using the Xcode IDE, it is much simple to reuse and adapt an existing one.

We can get one from the Titanium KitchenSink application, or we can use the one provided in this chapter's public GitHub repository once we have copied the `Settings.bundle` file into the `/platform/iphone` directory.

Inside Titanium Studio, we will expand the `Settings.bundle` branch and double-click on the `Root.plist` file. This will open the **Property List** editor provided by Xcode. The editor provides an easy way to modify properties files similar to this one.

We want our **Settings** window to have a group header labeled **Social Networks**, and underneath, one toggle Switch for each network. There are two setting identifiers (`facebook_preference` and `twitter_preference`). These identifiers will be used to retrieve their values through our code. To do this, we will modify the list in order to have the following configuration:

Key	Type	Value
iPhone Settings Schema	Dictionary	
Strings Filename	String	Root
Preferences Items	Array	
Item 0	Dictionary	
Type	String	Group
Title	String	Social Networks
Item 1	Dictionary	
Type	String	Toggle Switch
Title	String	Facebook
Identifier	String	facebook_preference
Default Value	Boolean	NO
Item 2	Dictionary	
Type	String	Toggle Switch
Title	String	Twitter
Identifier	String	twitter_preference
Default Value	Boolean	NO

This is all we need to do in order to create our **Settings** window (absolutely no code modification is necessary, which is a huge productivity gain while developing an application that has a lot of settings since an item can be a text field, a slider, a switch, a picker, or even a child window for other related settings).

Now to see if this works

With our `Property` file updated and saved, we can now take our **Settings** window for a spin. But before running an application, it is imperative that we clean the project using **Project | Clean...**

 The reason we need to clean the project every time we modify the **Settings** window is quite simple. During build time, configuration files such as this one are not always overwritten for performance purposes. So cleaning the project will force these same configuration files to be updated during the next build.

With a clean project, we can now run the application. Since Apple's guidelines expect application settings to be configurable through the native **Settings** app, we will have to quit our own application and go into the native iOS **Settings** app. By scrolling down, we see a **Unified Status** section, as shown in the following screenshot:

Once inside the **Unified Status** section, we can see our **Social Networks** group header with our toggle switches underneath:

If we were to change some of these settings and then come back later (even if we restart the simulator), these settings would remain the same since they are saved automatically on the device. This is another nice perk, considering we don't have to code it ourselves.

Settings for Android

Android shares a lot of similarities with iOS when it comes to settings management. Once again, the user interface is defined in a configuration file, and all the heavy lifting is done by the native platform. However, there is no visual editor here.

Since there is no need to start with an existing file, we will create a new file named `preferences.xml` and save it in the `/platform/android/res/xml` directory.

We will then edit the XML file and declare a `PreferenceScreen` with the `United Status` title as follows:

```
<?xml version="1.0" encoding="utf-8"?>
<PreferenceScreen
  xmlns:android="http://schemas.android.com/apk/res/android"
  android:title="United Status">
```

We will then create a category that will group our settings together. It will have `Preferences` as the title:

```
<PreferenceCategory android:title="Preferences">
```

Toggle switches are represented as checkboxes on Android. They act the same as with iOS, except that they have an extra property called `summary` that allows us to provide a short descriptive text to our users.

```
<CheckBoxPreference
  android:title="Facebook"
    android:defaultValue="false"
    android:summary=
  "Your messages will be published on your Facebook wall"
    android:key="facebook_preference" />
<CheckBoxPreference
  android:title="Twitter"
  android:defaultValue="false"
  android:summary=
  "Your messages will be published in your Twitter feed"
    android:key="twitter_preference" />
</PreferenceCategory>
</PreferenceScreen>
```

Once again, this is all we need to do to set our user interface on Android. You will notice that both the checkboxes use the same `key` identifiers as their iOS counterpart. Using the same keys will allow for the same code to access those values on both target platforms.

Where are those settings?

After cleaning the project (very important) and launching the application, you will notice that there is no way to access the **Settings** window. There is no dedicated app that centralizes application settings on Android. Each application must manage its own **Settings** window.

Since we do not want to modify the application's user interface just to access the **Settings** window (by adding a button), we will rely on a more conventional way for the Android platform.

Android menus

Menus have been present on Android since the earliest versions. For a long time, devices had a physical **Menu** button, which you could press, and then the different menus for the application would appear.

Since menus are only Android components, this entire code block will only be executed if we are running on the Android platform:

```
if (Ti.Platform.osname == "android") {
```

Android allows us to access a lot of native functionality through activities. Native menus cannot be attached to Titanium windows, but they can be attached to native activities.

According to the official Android API documentation, "an Activity is a single, focused thing that a user can do." In almost all cases, an activity is associated to only one window. So it makes perfect sense to assume that the `currentActivity` constant is associated with our main window. This deduction is easy because our application has only one window. But for larger applications, there are ways that we can use to identify them if needed.

We will retrieve our main window's activity as follows:

```
var activity = Ti.Android.currentActivity;
```

We will then call the `onCreateOptionsMenu` function provided by the `activity` object. We will retrieve the activity's (empty) menu and add a new menu item as follows:

```
activity.onCreateOptionsMenu = function(e) {
  var menu = e.menu;
  var menuItem = menu.add({
    title: "Settings"
  });
```

We will then assign an icon to our menu using the platform's default icon for the **Settings** menu using a system constant. Using the default icons is usually recommended in order to provide a more coherent user experience throughout different applications:

```
menuItem.setIcon(
  Ti.Android.R.drawable.ic_menu_preferences);
```

Finally, we will add an event handler on the `click` event for our newly created menu item. Inside this event handler, we will call the generic Android `openPreferences` function in order to display the **Settings** window:

```
menuItem.addEventListener("click", function(e) {
  Ti.UI.Android.openPreferences();
  });
  };
}
```

There is a rule on Android that basically states that every activity must be declared in the Android Manifest. While building a Titanium application for Android, all the activities are declared automatically. But since the preferences activity (the **Settings** window) is not a Titanium window, it will never be declared.

This is why we will declare it in our `tiapp.xml` configuration file by adding `TiPreferencesActivity` to the `android` section as follows:

```
<android xmlns:android="http://schemas.android.com/apk/res/android">
  <activity
  android:name=
  "ti.modules.titanium.ui.android.TiPreferencesActivity"/>
</android>
```

As of Android 4.0, devices got rid of the hardware Menu Button altogether. But menus are still available from the bottom bar; you will notice three little dots at the right. Touching these will display the menu.

Let's give it a run!

Now that everything is in place, we can now run our application in Android. Once the application is loaded, we will hit the **Menu** button, and we should see our **Settings** menu appear at the bottom of the screen, as shown in the following screenshot:

When we tap the menu, the native **Settings** window is displayed. Although it looks different, it has the same features set as its iOS counterpart, as shown in the following screenshot:

The settings are changed, then what?

Now that our users are able to modify their application's settings at their will, we must reflect those same changes in our application. We will create a dedicated function that will load the settings and apply them to the application using the following code:

```
function loadSettings() {
```

The settings can be retrieved just as we did for any other property using the keys we defined while creating our **Settings** window earlier. So we will just retrieve the values and store them in variables for later use using the getBool function. In case no settings have been assigned (the application was just installed), we will pass false as a default value in the second parameter:

```
var fb = Ti.App.Properties.getBool('facebook_preference', false);
var tw = Ti.App.Properties.getBool('twitter_preference', false);
```

We will then invoke the toggle function as follows:

```
    toggleFacebook(fb);
    toggleTwitter(tw);
}
```

It is also important to call this new function right before the main window is opened:

```
loadSettings();
```

It goes both ways

Since changes made in the Settings section are reflected in the application, we will apply those same rules when our users make changes to their settings from inside the application. A very simple way to do this is to update the properties when the user toggles a social network just by adding one line for each toggle function as follows:

```
function toggleFacebook(isActive) {
    ...
    Ti.App.Properties.setBool('facebook_preference', isActive);
}

function toggleTwitter(isActive) {
    ...
    Ti.App.Properties.setBool('twitter_preference', isActive);
}
```

Summary

In this chapter, we learned how to create server-side applications on two popular social networking websites. We learned how to interact with those networks in terms of API and authentication. We also learned how to handle device rotation as well as use the native platform setting Windows. Finally, we covered Titanium menus and activities.

In the next chapter, we will be delving into Geolocation as well as a much more powerful alternative to the TableView component.

9
Marvels of the World around Us

In this chapter, we will be developing an application that shows photos taken around the users location. The application will use Geolocation to determine the device's current location. It will then use the same coordinates to query a web service and retrieve photos that were shot near those coordinates. Our user will then be able to select a particular photo from the list and display it on the screen. We will also give our user the ability to add a photo they like to their photo galleries.

By the end of this chapter, you will have covered the following concepts:

- Getting the device's coordinates
- Retrieving photos based on nearby locations
- Displaying the pictures using the new `ListView` component
- Saving the desired pictures to the device's Image Gallery

Creating our project

As with our previous projects, we need to set up a new project for our application. To do this, select the **File | New | Mobile Project** menu from Titanium Studio and fill out the **Wizard** forms with the following information:

Field	Value to enter
Project Template	Classic, default project
Project Name	PhotoSurrounder
Location	You can either:
	• Create the project in your current workspace directory by checking the **Use Default Location** checkbox
	• Create the project in a location of your choice
App Id	com.packtpub.hotshot.photosurrounder
Company/Personal URL	http://www.packtpub.com
Titanium SDK Version	By default, the wizard will select the latest version of the Titanium SDK. This is recommended (as of now, we are using version 3.1.3 GA)
Deployment Targets	Check **iPhone**, **iPad**, and **Android**
Cloud Settings	Uncheck the **Cloud-enable this application** checkbox

Project creation is covered in more extensive detail in *Chapter 1, Stopwatch (with Lap Counter)*. So feel free to refer to this section if you want more information regarding project creation.

The main window

As usual, we want to look before we leap. This means that we have to clearly define our application's user interface before we write a single line of code. Our main window will comprise a container View at the top to hold a label and a button used to refresh the list. The rest of the window will be occupied by a ListView component that will act pretty similar to a TableView component. Each row from the list will comprise an ImageView component for the photo itself and two labels used to display the photo's related information.

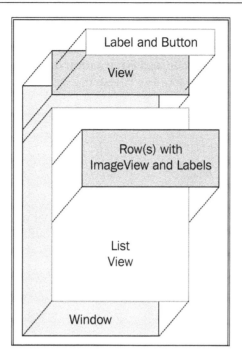

Let's dig in the right way

With our UI mockup in hand, we can move on to the implementation. We will create a new file named ApplicationWindow.js in our Resources directory. This new file will contain all the code related to our main window.

We will wrap all our code into a single function called ApplicationWindow:

```
function ApplicationWindow() {
```

We will then create our window object and store its reference into a variable named self for later use:

```
var self = Ti.UI.createWindow({
  backgroundColor: '#fffbd0'
});
```

The header

We will then create the `header` View that will sit at the very top of the screen. It will span the whole width of the screen and have a nice, dark orange color. Once created, we will keep its reference in the `header` variable:

```
var header = Ti.UI.createView({
  top: 0,
  width: Ti.UI.FILL,
  height: 50,
  backgroundColor: '#c13100'
});
```

We will then add a title to our header using a white label with a bigger and bolder font size. Since we won't be modifying this label from our code, we will be adding it to our header in the same statement:

```
header.add(Ti.UI.createLabel({
  text: 'Photo Surrounder',
  color: '#fff',
  left: 10,
  font:{
    fontSize: '22sp',
    fontWeight: 'bold'
  }
}));
```

The last component of our header is our **Refresh** button, located on the top right of the screen. We will be using an ImageView component since we want it represented by an icon on the screen. We will keep its reference in the `btnRefresh` variable and immediately add it to our header as follows:

```
var btnRefresh = Ti.UI.createImageView({
  image: 'refresh.png',
  right: 3
});
header.add(btnRefresh);
```

To finish with our header object, we will add it to the main window; otherwise it will not be displayed:

```
self.add(header);
```

The ListView component

To display the photos around our location, we will be using a ListView component. The ListView component is a more recent component and was designed to be a replacement for the TableView component. At the end, they both provide the same functionality; which is to display data in a scrollable list of rows. But the ListView component is designed to perform better while dealing with a large number of rows; something the TableView component had trouble doing smoothly when it had to display a high number of rows with complex layouts.

One of the major differences while using ListView components is that there is a clear separation between the visual structure of each row and the data represented underneath. This has the effect of limiting direct access to individual rows, meaning that you wouldn't be able to access, for example, an ImageView component from a specific row.

Instead, you define templates that will define how your data will be displayed in the row (similar to what you would do for any control on a regular view). You can define as many templates as you need and each row (or a given set of rows) can have its own unique UI structure.

The template

Each of our rows will share the same UI structure; they have an ImageView component on the left with the photo's thumbnail, a label next to it with the photo's title, and a second label that will be display the photo's geographic information (where the photo was actually taken). A ListView template is basically a regular JSON object containing the visual properties as well as an array of components that will be used to create the rows of UI (they are called `childTemplates`).

We will now go ahead and create our template and keep its reference into a variable named `photoTemplate`. Using the `properties` attribute, we will give each row a specific height and we will also display a detailed indicator on the right side of the row (to indicate that selecting the row will display the item's details).

```
var photoTemplate = {
  properties: {
    height: 60,
    accessoryType: Ti.UI.LIST_ACCESSORY_TYPE_DETAIL
  },
```

We will then create a new array containing all the items required to construct our row. This new array will then be assigned to the `childTemplates` property.

The first item from our template will represent the photo's thumbnail. We will set its type to ImageView to display the thumbnail. We will then give it a technical ID (bindId) that we will use to access it later in our code; we will use thumb for the value of this ID. We will also add some additional properties to position our row:

```
childTemplates: [ {
  type: 'Ti.UI.ImageView',
  bindId: 'thumb',
  properties: {
    left: 0,
    width: 45
  }
},
```

The second item will be of the type label and will be used to represent the photo's title. We will use rowtitle as its bindId property for later reference and will give it an orange color. We will also position it using properties:

```
{
  type: 'Ti.UI.Label',
  bindId: 'rowtitle',
  properties: {
    left: 48,
    top: 1,
    color: '#cc6600'
  }
},
```

The third and the last item will also be labels and will be used to display the photo's geographic information. We will use coordinates for its bindId property and apply style properties just as we would do for a regular label. In fact, every property supported by the original label object can be used here:

```
{
  type: 'Ti.UI.Label',
  bindId: 'coordinates',
  properties: {
    left: 48,
    bottom: 2,
    width: '75%',
    textAlign: Ti.UI.TEXT_ALIGNMENT_CENTER,
    color: '#fff',
    backgroundColor: '#ff9900',
    font: {
      size: '6sp'
    }
  }
} ]
};
```

Creating our ListView object

Now that we know how our data will be presented on the screen, we will now create our ListView object. The most important thing to do here is to create a link between our Template and the ListView components. This is done using the `defaultItemTemplate` property. But there is a catch; this specific property can only accept a string, not a template object.

This is why we will create an object that will act as a map between our `photoTemplate` object and a string of our choice. In our case, we will use `'photo'` to keep it simple and store this key-value relation in the `templates` property.

Once the mapping is done, we will set the `defaultItemTemplate` property to the value of the `'photo'` character string we just declared:

```
var listView = Ti.UI.createListView({
  top: 50,
  templates : {
    'photo': photoTemplate
  },
  defaultItemTemplate : 'photo'
});
```

> While it might appear a little redundant to create a map between a string and a template, we must keep in mind that a ListView component can have multiple templates and this map will be the only way to select which template will be used while rendering a specific row or a set of rows.

Once created, we will add our newly created ListView component to our window:

```
self.add(listView);
```

Wrapping it up

With all our components added to our main window, our function will return the `self` object (that is, the window itself). This means when `ApplicationWindow` is invoked, it will return our window:

```
  return self;
};
```

As with every CommonJS module, we will export the function in order to be able to load it using the `require` function:

```
module.exports = ApplicationWindow;
```

Calling our main window

We will now need to invoke our main application window every time the application starts. We will clear all the default content from the app.js file and replace it with a simple self-calling function. This function will load ApplicationWindow object's CommonJS module, create a new window, and then open that window:

```
(function() {
  var ApplicationWindow = require('ApplicationWindow');
  var win = new ApplicationWindow();

  win.open();
})();
```

Testing our work so far

Let's take our application for a spin by clicking on the **Run** button from the **App Explorer** tab.

Here is what we see on our very first run:

While this is not much yet, this allows us to make sure whether our main window is functional. Of course, there are no photos in the ListView component, and we will remedy that shortly.

Getting the device's location

In order to search photos that were taken around the device's location, it is imperative that we determine where this location is. Titanium provides us with a Geolocation module that is provided with the SDK.

The module provides the following two main families of features:

- Location services that can be used to determine the location of the device.
- Geocoding and reverse geocoding that can be used to convert geographic coordinates into street addresses. The reverse is also possible, meaning that we can convert physical addresses into geographic coordinates.

 Be always thoughtful of power consumption when dealing with location services. Many factors such as accuracy and update frequency can have a huge impact on the device's battery life, if it is not used properly.

How does that translate into code?

Inside our `ApplicationWindow` function, after defining the user interface, we will add our Geolocation code.

On iOS specifically, the user must approve the use of the location services. We will specify the text from this confirmation dialog using the `purpose` property:

```
Ti.Geolocation.purpose = "Location will be used for photosearch";
```

We will create a `refreshData` function that will be called every time we want the location of the device to be updated. In our case, this function will be called automatically when the application starts as well as when the user clicks on the refresh button:

```
self.refreshData = function() {
```

We will then call the `getCurrentPosition` function, which expects a callback function as the parameter:

```
Ti.Geolocation.getCurrentPosition(function(e) {
```

The function's name is pretty self-explanatory and when it succeeds, it returns quite a lot of location-related information, which are as follows:

- The accuracy (in meters)
- The altitude and its accuracy (in meters)
- The compass heading (in degrees)
- The latitude and longitude (in decimal degrees)
- The current speed (in meters per second)
- The timestamp (in milliseconds)

In case of error, we will log the error message to the console and show it to our user using an alert. We will invoke the return statement, so that the code of this function stops there:

```
if (!e.success || e.error) {
  Ti.API.error(JSON.stringify(e.error));
  alert('error ' + JSON.stringify(e.error));

  return;
}
```

In case of success, we log the returned coordinates' longitude and latitude to the console:

```
    Ti.API.info('Geolocation:'
        + ' long ' + e.coords.longitude
        + ' lat ' + e.coords.latitude);
  });
}
```

In order for the `refreshData` function to be called automatically at startup, we will add an event listener for when the main window is opened:

```
self.addEventListener('open', function() {
  self.refreshData();
});
```

How can we test this?

To test a feature such as Geolocation using the emulator, we will require some adjustments. Since neither of them have an embedded GPS or compass, we will have to simulate our own location by using either of the following options:

Using the iOS simulator

Once the simulator is started, we can set its location by navigating to the **Debug | Location** menu. From there, we can choose from different preset locations such as Apple headquarters or an Apple retail store. We can also choose **City** or **Bicycle Run**, and **Freeway Drive** to simulate a moving location (in Cupertino, of course).

We can even set our own custom location using longitude and latitude.

Finally, we can deactivate the location service using the **None** menu if we want to test how our application would react when it cannot determine the device's location.

Using the Android emulator

The Android emulator provides a similar feature; but it must be accessed using a tool called the Android Debug Monitor. It is a free graphical tool that comes with the Android SDK. It is a very powerful program that allows developers to access logs, memory usage, processes, services, performance metrics, and many other things that go on under the hood. The Debug Monitor also gives us control of the device's features such as network, battery level, incoming calls or messages, and location.

 The Android Debug Monitor can be used with both an emulator and a physically connected device. This becomes very useful while debugging real-life scenarios on a specific model of device (which can be legion on the Android platform).

To access it we will go to the `/tools` directory from our Android SDK Installation and launch the `monitor` program.

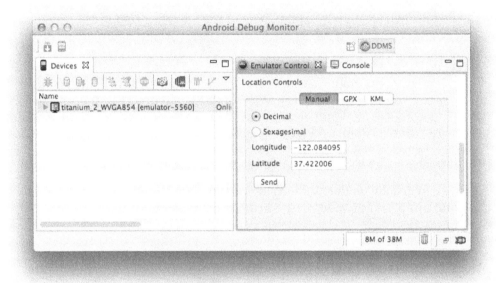

Once the program is launched, we will select our emulator (or device) from the **Devices** view (tabs are called Views just like in Titanium Studio). With our emulator selected, we can have access the **Emulator Control** view. Inside this view, we will have access to the **Location Controls** section, which will allow us to enter our desired longitude and latitude coordinates.

Validating our test

Now that we know how to use Location Services on our testing platform(s), we can run our application just like we would on a normal device. When we launch our application, we should see a message matching this one from our console. (The coordinates may be different depending on your configuration):

```
[INFO]  Geolocation: long -122.40641784667969 lat 37.78583526611328
```

Since Gelocation is technically contained in its own Titanium Module, you might need to clean the project (using the **Project | Clean...** menu) in order to force a full rebuild. This is usually a necessary step when using new Modules since Titanium doesn't do a full rebuild every single time (for speed purposes).

Photos from the Web to our application

We now have an application that is aware of its position on the globe and we will now capitalize on that feature by searching for photos taken within a 5-km radius. All these photos will be (of course) retrieved from the web and then added to our ListView component.

There is a web service for that...

There are many online photo websites out there and Flickr is still a very popular service that is widely used today. Moreover, it provides an API that will allow us to retrieve photos using geographic data.

Getting our API key

Similar to what we did when we interacted with social networks, we need to identify our application to the Flickr service. To do this, all we need is a Flickr account. It's basically a Yahoo ID, but the service now allows users to log in using their Facebook or Google credentials.

Once logged in, we will create our new application at the following URL:

```
http://www.flickr.com/services/apps/create/
```

From there, we will request an API key for our application. This key will have to be used every time we want to access the web service down the road. The form will require us to enter a name as well as a mandatory short description for our application.

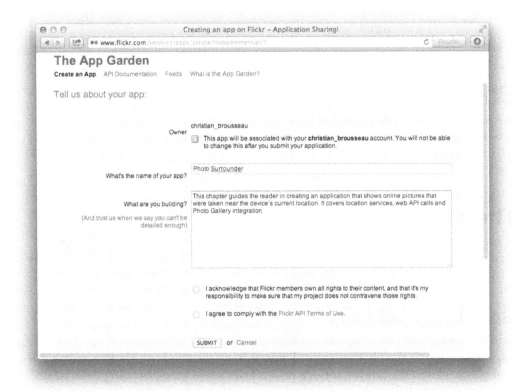

Once created, we will be issued with our API key.

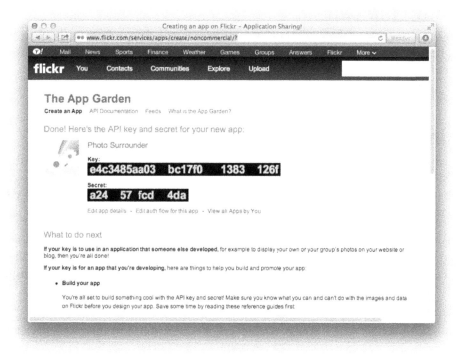

Remembering that key

Since we want to easily modify our API key without modifying the source code itself (we might be using a different API key for development than the one used for production), we will add the following code into our `tiapp.xml` file:

```
<property name="flickr.key">[YOUR_FLICKR_API_KEY]</property>
```

Once the property has been made available, we will create a variable that will store the key's value named `FLICKR_KEY` right at the beginning of our `ApplicationWindow` function:

```
var FLICKR_KEY = Ti.App.Properties.getString('flickr.key');
```

Defining the call

Calling a web service is basically a standard HTTP call. The only difference in this case is that we will expect a JSON-based response instead of HTML. To do this, we will create a new HTTP client named xhr as follows:

```
var xhr = Titanium.Network.createHTTPClient();
```

Making the call

The Flickr API provides us with several methods to provide access to almost the entire site's content. Our application will want to search every photo with geographic data that has been taken near the device's coordinates. We also want our query to return the image's URL as well as the thumbnail URL.

The flickr.photos.search method provides all those features and more. This method has 34 parameters (all optional) at the time of publication and we can't cover all of them in this chapter. But here are some of the parameters we will be using:

Parameter	Description
method (required)	This parameter specifies which API method we will call (flickr.photos.search in our case).
api_key (required)	This parameter specifies our API key.
has_geo	This parameter is used to specify that we want our search to cover only the photos with geographic data.
lat	This parameter specifies the latitude.
lon	This parameter specifies the longitude.
extras	This parameter represents a comma-delimited list of extra information to fetch for each returned record. This will be used to specify that we want to retrieve the photo's URL (url_n) and the photo's thumbnail URL (url_t). All supported fields are available from the API documentation.
format	This parameter specifies the returned format we are expecting. It can be XML, JSON, JSONP (basically JSON with a callback function), or even PHP serial.

We will now update our refreshData function in order to call the web service every time the device's location is determined.

```
self.refreshData = function() {
  Titanium.Geolocation.getCurrentPosition(function(e) {
    ...
```

We will then open our HTTP client request by constructing the API URL using all the required parameters we listed previously and then send the request to the server:

```
xhr.open('GET', 'http://api.flickr.com/services/rest/'
    + '?method=flickr.photos.search'
    + '&api_key=' + FLICKR_KEY
    + '&has_geo=true'
    + '&lat=' + e.coords.latitude
    + '&lon=' + e.coords.longitude
    + '&extras=geo%2Curl_t%2Curl_n'
    + '&format=json'
    + '&nojsoncallback=1');
xhr.send();
```

Also, we want to clear all data that may already be present in our `listView` object. The easiest way to achieve this will be to delete the very first section from the list. It is similar to the `TableView` component. All the data is contained in at least one section by default (some applications may require additional sections to display data):

```
    listView.deleteSectionAt(0);
  });
}
```

Handling the response

Until this point, we have successfully called the Flickr API using our HTTP client. We will now handle the response from the service by assigning the `onload` property with a callback function that will be executed upon a successful response from the service:

```
xhr.onload = function() {
```

We will declare all the variables we need to store the information contained in the response. The most notable one is the `json` variable containing the whole response transformed as a JavaScript object. The `jsonImages` array will contain the photo collection itself. Other notable variables are two arrays that we will use to store the photo objects and the thumbnail URLs:

```
var json = JSON.parse(this.responseText),
  jsonImages = json.photos.photo,
  image,
  images = [],
  preview = [];
```

We will now loop through every photo returned by the service and populate both arrays using the following code:

```
for (index in jsonImages) {
  image = jsonImages[index];
  images[index] = image;
  preview[index] = image.url_t;
}
```

With our arrays filled with all the information we need, we will now fill our ListView data. For this, we will loop through our `images` array containing our photos, as follows:

```
var data = [];

for (var i = 0; i < images.length; i++) {
```

We will then add a row for each photo. This is where it is important to remember how we defined our template earlier in this chapter because we will be using the same `bindId` to map our data to our template elements.

The `rowtitle` template component from our template is a label, so we will assign its `text` property with the photo's title:

```
data.push({
  rowtitle: {
    text: images[i].title
  },
```

The `thumb` component is an ImageView component. We will then assign its `image` property with the thumbnail's URL:

```
thumb: {
  image: preview[i]
},
```

The `coordinates` component is another label. We will assign its `text` property with the photo's geographic coordinates:

```
coordinates: {
  text: images[i].longitude + ', '+ images[i].latitude
},
```

We will also set an additional property to our `ListDataItem` object (the actual class name for a `ListView` row item) to help us determine which row was clicked when our user interacted with the list. This is done using the `itemId` property. Since this property only accepts a string (on Android), we will transform our photo JavaScript object into a string, which will save us time:

```
        properties: {
          itemId: JSON.stringify(images[i])
        }
      });
    }
```

 Contrary to the `TableView` component, we cannot simply assign a photo object to our row, meaning that we must use some sort of identifier for each row and then retrieve the actual data object using that same identifier. This is mainly due to the fact that ListView components have a clear separation between the data to be displayed from the individual view structure of each row, which limits direct access to each individual row. Even though this may require some extra work on the developer's part, the performance gains are definitely worth it.

Since all the ListView data must be contained into a `ListSection` object, we will simply create one and append it to our `listView`:

```
    var section = Ti.UI.createListSection({
      items : data
    });

    listView.appendSection(section);
  };
```

Having a second look at our main window

Now if we run the application again, we will see our ListView component being populated when the application starts.

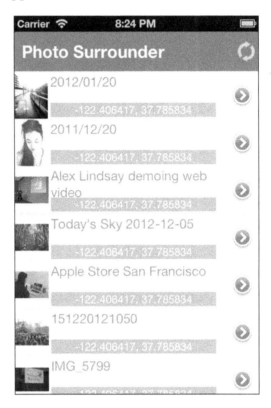

Now, even though there are many rows in the list, we can see that the scrolling is very fast and smooth, which would have been difficult to replicate using the TableView component.

 Be sure to use a well-known location while testing location-aware data services (usually a big city or a tourist prone area). This will ensure that you have sample data that is representative enough to test most of your use cases.

The photo viewer window

The photo search part of our application is now behind us and we will now move on to our second window that will be used to view the selected photo when the user clicks on a specific row.

The user interface for this window is quite straightforward. We will have a header **View** sitting at the top of the screen that will contain both **Label** and **Buttons** for the photo's title.

The rest of the window will be entirely occupied by an **Image View** component that will display the photo:

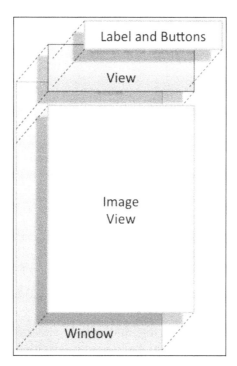

On with the code

For the sake of clarity, we will keep all the code related with this window in its own file. We will create a new file named `PhotoViewerWindow.js` in our `Resources` directory.

In this newly created file, we will create a function named `PhotoViewerWindow` that will encapsulate everything related to our window. This function will take a single argument (the selected photo) and since the parameter `itemId` is a string, we will parse it into a JavaScript object:

```
function PhotoViewerWindow(itemId) {
  var image = JSON.parse(itemId);
```

We will then create our window with a black background and make sure that the navigation bar is visible (not hidden). We will keep its reference in a variable named `self`:

```
var self = Ti.UI.createWindow({
  backgroundColor: '#000',
  navBarHidden: false
});
```

Showing the navigation bar is a simple hack to ensure that the window is "heavyweight", meaning that it will be associated with an Android activity.

We will create our `header` View that will be placed at the top, will span the whole width of the screen, and will have a dark orange background. Since we want to make sure it is always on top of any other UI component, we will give it a `zIndex` of `10`:

```
var header = Ti.UI.createView({
  backgroundColor: '#c13100',
  zIndex: 10,
  top: 0,
  width: Ti.UI.FILL,
  height: 50
});
```

We will add a label to our `header` View that will display the `image` object's `title` property. It will be white and will have a bigger and bolder font. Also, by not setting any positional properties, we will ensure that the label is centered into its parent View.

```
header.add(Ti.UI.createLabel({
  text: image.title,
  color: '#fff',
  font:{
    fontSize: '18sp',
    fontWeight: 'bold'
  }
}));
```

Our `header` View will contain a back button at the very left side to allow our users to go back to the list and select another photo they desire. We will use an ImageViewin order to use an icon, not forgetting to add it to our `header` View.

```
var btnBack = Ti.UI.createImageView({
  image: 'back.png',
  left: 3
});
header.add(btnBack);
```

Our header View will also contain a **Save** button to allow our users to save a photo they like to their device. It will be located at the top-right corner of the screen. We will also add it to our `header` View:

```
var btnSave = Ti.UI.createImageView({
  image: 'save.png',
  right: 3
});
header.add(btnSave);
```

Let's not forget to add the `header` View to our window:

```
self.add(header);
```

The last UI component required for our window is an ImageView component that will be used to display the image that was passed to our window as parameter. It will occupy the whole screen and we will set the photo's Flickr URL as its `image` property:

```
var photoView = Ti.UI.createImageView({
  width: '100%',
  height: '100%',
  image: image.url_n
});
self.add(photoView);
```

Finally, our `PhotoViewerWindow` function will return our newly created window and we will export the function so that it is accessible using the `require` function:

```
  return self;
}

module.exports = PhotoViewerWindow;
```

Returning to our main window

While selecting an image from the list will take our users to our photo viewer window; but we must provide them with a way to return back to the list. To achieve this, we will simply add an event handler on the click event of our btnBack button. Once this handler is triggered, we will close the window, thus navigating back to the main window:

```
btnBack.addEventListener('click', function() {
  self.close();
});
```

Connecting the wires

We now have two completely functional windows in our application. But we need to create some sort of navigation bridge in order for our users to navigate from one window to the next. While we are at it, we will give users the ability to manually refresh the list as their geographic location might change while using the application.

Selecting a photo

The very first thing we will do is to load the PhotoViewerWindow module from our main window. To do this, we will add the following line at the very top of the ApplicationWindow.js file:

```
var PhotoViewerWindow = require('PhotoViewerWindow');
```

To open the photo viewer window when the user selects a photo from the list; we will add an event handler to our ListView object. When the itemclick event is triggered, we will create a new instance of the PhotoViewerWindow class by passing it the itemId event from the selected row. We will then open our newly created window:

```
listView.addEventListener('itemclick', function(e){
  var photoWin = new PhotoViewerWindow(e.itemId);

  photoWin.open();
});
```

The Refresh button

We will allow our users to manually determine the device's location, query the web service, and refresh the list. This can prove useful when the user's location changes while using the application.

Luckily for us, the `refreshData` function does all these operations already. So, we will leverage this same function in the `click` event listener for the **Refresh** button we created earlier:

```
btnRefresh.addEventListener('click', function() {
  self.refreshData();
});
```

Testing out the navigation

We will now test our application one more time to test the photo viewer's functionality as well as to make sure that the navigation works as intended. Once the application is started and the ListView object is populated with photos taken near the device's location; if we simply tap one from the list we will be taken to our next window:

Once this second window is displayed, we can return to the list by tapping the back icon at top-right of the screen.

Photo gallery integration

The last segment of this chapter will allow our users to save any picture they like from the application into their device's photo gallery. Both iOS and Android provide their own brand of photo gallery so that each can be addressed specifically.

The very first thing we will do is to add an event handler when the user clicks on the **Save** button:

```
btnSave.addEventListener('click', function() {
```

Android photo gallery integration

In this handler, we will isolate the code specific to Android in a conditional block:

```
if (Ti.Android) {
```

We first need to determine a directory where our photo will be saved on the device. In our case, we will be using the `externalStorageDirectory` property that is the path to the device's storage (such as an SD card).

```
var tempDir = Ti.Filesystem.externalStorageDirectory;
```

 It is recommended to check if the location is available using the `isExternalStoragePresent` function before accessing it. This will prevent errors in cases where external storage is not available on the device.

We will then create a new file using the `getFile` function. This function expects two parameters; the first one is the location where the photo will be saved and the second one is the name of the file. Since we want each photo to have a different name, we will retrieve the device's current time (in milliseconds):

```
var newFile =Ti.Filesystem.getFile(tempDir, new Date().getTime() +
'.jpg');
```

Now we will need to get the ImageView object's content into our newly created file. We will then capture an image of the rendered view itself by using the `toImage` function and save its content into a variable named `f`, for later use:

```
var f = photoView.toImage();
```

 The image object contains many properties such as dimensions, mime type, native path, resolution, and the binary data of the image.

We will then write the image's actual data (contained in the `media` property) into our file as follows:

```
newFile.write(f.media);
```

Now that our photo is properly saved on our device, we must tell our device to scan the new media so that the photo gallery knows that it actually exists. The `Titanium.Media.Android` module provides a function named `scanMediaFiles` that does just that. This function requires an array of the paths to be scanned. In our case, it is the photo's native path on the device. It also allows for two extra parameters for specific MIME types as well as a callback function (if needed):

```
Ti.Media.Android.scanMediaFiles([newFile.nativePath], null,
function(e) { });
```

iOS photo gallery integration

In our iOS-specific block (the `else` case), we will invoke the `saveToPhotoGallery` function provided by the `Titanium.Media` submodule. This function takes a single parameter representing the media to be saved in the photo gallery. The media can be physical, that is from the device's filesystem, or a blob object (a container of binary data).

Luckily, the ImageView object provides us with a `toBlob` function that returns its image content in binary format. So, we will simply call this function and pass it as the parameter while saving the image to the photo gallery:

```
} else {
  Ti.Media.saveToPhotoGallery(photoView.toBlob());
}
```

Once the photo is saved to the target platform's gallery, we will show an alert dialog window to inform the user that the image is now available in their photo gallery:

```
Ti.UI.createAlertDialog({
  title: 'Photo Gallery',
  message: 'Photo added to your photo gallery'
}).show();
});
```

One final run

Now, the time has come for our very last test run. Once the application is started, we will select a photo from the available list. Once the photo is displayed, we will tap the **Save** button at the top-right corner of the screen. This will save the photo to the device's photo gallery and we will be informed by an alert dialog.

Summary

This chapter details how to create an application that relies on the device's geographic data. We learned how to retrieve that location and then use it to invoke a web service to retrieve photos based on that same location. We also learned how to use the new `ListView` UI component, which has far better performance than regular `TableView` components. Finally, we leveraged the photo gallery integration to allow our users to save the photos they like on their device.

In our next chapter, we will leverage Appcelerator Cloud Services to share data with all our users.

10
Worldwide Marco Polo

In this chapter, we will be developing a Marco Polo-like game that uses the world as the playground. Players will have the ability to check in wherever they are in the world and see other players' locations using the device's Geolocation feature (as well as yours) on a map view. Since we want all of the locations shared among all the players, we will rely on **Appcelerator Cloud Services(ACS)** to store and share our geographic information. This chapter will also cover how to use map views and add annotations to the map.

By the end of this chapter, you will have learned the following concepts:

- Integrating a Titanium project with ACS
- Using tabs in our application
- Using more advanced Geolocation features
- Using the MapView component
- Adding annotations to a map
- Using the latest Google Maps v2 for Android

We know the drill

We will create a new project for our application, just as we created the previous ones. To do this, select **File | New | Titanium Mobile Project** from Titanium Studio and fill out the wizard forms with the following information:

Field	Value to enter
Project Template	`Classic`, the default one
Project Name	`Worldwide Marco Polo`
Location	You can follow either one of the ensuing steps:
	• Create the project in your current workspace directory by checking the **Use Default Location** checkbox
	• Create the project in a location of your choice
App Id	`com.packtpub.hotshot.marcopolo`
Company/Personal URL	`http://www.packtpub.com`
Titanium SDK Version	By default, the wizard will select the latest version of the Titanium SDK; this is recommended (at the time of writing this book, we are using version 3.1.3 GA)
Deployment Targets	Check **iPhone**, **iPad**, and **Android**
Cloud Settings	This time, we will check the **Cloud-enable this application** checkbox

Creating a project is covered more extensively in *Chapter 1, Stopwatch (with Lap Counter)*, so feel free to refer to this section if you want more information regarding project creation.

Let's see what we got here

By checking the **Cloud-enable this application** checkbox, the project creation wizard adds a few extra things to our `tiapp.xml` file. The first thing that will be added is a reference to the `ti.cloud` module, as it appears in the following code:

```
<modules>
  <module platform="commonjs">ti.cloud</module>
</modules>
```

The wizard also created six string properties acting as keys required to interact with **Appcelerator Cloud Services** (**ACS**). Those keys come in two different sets: one we will use during our development phase and another one we will use once our application is in production (distributed to customers).

Property Name	Description
`acs-api-key-[ENVIRONMENT]`	Our application needs its own key in order to prove it is allowed to interact with ACS. This will keep our data secure by preventing anyone making requests to ACS impersonating our application.
`acs-oauth-key-[ENVIRONMENT]` `acs-oauth-secret-[ENVIRONMENT]`	ACS also supports the `OAuth` authentication, which is considered more secure. Here, a key and a secret are used to sign every single request made by our application. When the ACS server receives the request, the secret is used along with the data sent in the request to calculate another signature. If the sent signature and calculated signature match, the request will be processed.

It is also possible to add the ACS functionality to an existing Titanium project using Titanium Studio. To do this, simply open the project's `tiapp.xml` and switch to the **Overview** tab. From there, click on the **Enable...** button located next to the **Enable Cloud Services:** label, as shown in the following screenshot:

Our tabbed user interface

Our application has two main themes, Marco and Polo as the name implies. The **Marco Window** will be used to locate other players and display them on a **Map View** using annotations (pins). When our player selects a pin, the annotation's details will be displayed; in this case, it is the other player's name.

The **Polo Window** will give players the ability to register their location into the Cloud for other players to see (on their map). The window will have a label and a text field at the top asking players to enter their name. It will also have a big circular view in the middle so that the players can click on it when they want to update their location. Finally, it will have a label at the bottom that will indicate the status of what is going on in the background.

Instead of displaying one window at a time and opening and closing each of them via code, we will delegate this navigation to the **Tab Group** component. This component will wrap each window into a tab and give users the ability to move from one window to another simply by selecting the tab of their choice.

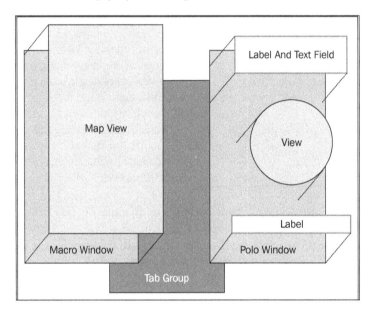

Getting this show on the road

Now that we have a clear vision, we can dive into the code right away. Our first order of business will be to create a **Tab Group** component and make sure our navigation works as intended. This will avoid surprises later on when the application is a lot more complex.

The chicken and the egg

As we covered earlier, a `TabGroup` component contains windows and takes care of the navigation from one window to another. Of course, we need to have windows to begin with. Therefore, we will create all of our windows from the get-go.

To have a clear separation between each window, we will create a dedicated file named `MarcoWindow.js` in our `Resources` directory. Inside this newly created file, we will create a function named `MarcoWindow` that will return a new window. This window will have a title, a white background, and a dark red colored navigation bar. We will finally export the function since we want to invoke it using the `require` function later:

```
function MarcoWindow() {
  var self = Ti.UI.createWindow({
    title: 'Marco',
    backgroundColor: '#fff',
    barColor: '#8C001a'
  });

  return self;
}


module.exports = MarcoWindow;
```

Our Polo window will also be contained in its source file. We will create a new file named `PoloWindow.js` in our `Resource` directory. Just as before, our function will return a new window with a light yellow background and share the same dark red color for its navigation bar. We will then export our new function at the end of the file as follows:

```
function PoloWindow() {
  var self = Ti.UI.createWindow({
    title: 'Polo',
    backgroundColor: '#f6fa9c',
    barColor: '#8C001a'
  });

  return self;
}
module.exports = PoloWindow;
```

Although there are no controls for these windows yet, the preceding code will at least give us enough to use them within our `TabGroup` component. Also, since they each have a different background color, we will be able to verify which window is assigned to each tab.

Using those windows

Having created our two windows, we can now proceed to create our group of tabs. We will now edit our app.js file and replace its entire content with our code. As usual, we will create a self-invoking function to make sure all our application's variables and objects have a unique namespace:

```
(function() {
```

We will then create our group of tabs and store their reference in a variable named tabs. This will act as the main container for our entire application. In fact, our application will not have a main window similar to the ones we covered previously. The TabGroup component will handle everything:

```
var tabs = Ti.UI.createTabGroup();
```

We will then invoke every single window in advance in order to add them to our tabs later on. Each window will have its reference stored in its own variable for later use:

```
var winMarco = require('MarcoWindow')();
var winPolo = require('PoloWindow')();
```

 Notice how we created our windows using a simple one-line code. To do this, all we have to do is add an extra set to parentheses. This will basically call the function assigned to the module.exports property from our CommonJS module. In our case, each of those functions returns a new window.

We will now create our two tabs using the createTab function. Each tab will have a specific title attribute as well as an icon image. But more importantly, we will assign the window property of the window we want to display when the tab is selected:

```
var tabMarco = Ti.UI.createTab({
  title: 'Marco',
  icon: 'marco.png',
  window: winMarco
});

var tabPolo = Ti.UI.createTab({
  title: 'Polo',
  icon: 'polo.png',
  window: winPolo
});
```

Once our tabs have been created and each of them has a window assigned to it, we will add all of them to our `TabGroup` component using the `addTab` function. It is important to mention that the order in which they are added to the group directly reflects the order they will be presented on the screen:

```
tabs.addTab(tabMarco);
tabs.addTab(tabPolo);
```

As we mentioned earlier, the `TabGroup` component acts as our main container; we will call the `open` function to display it on the screen without forgetting to close and invoke our self-containing function we created at the very beginning:

```
    tabs.open();
})();
```

Running our application for the first time

We now have enough code to take our application for a spin. If we click on the **Run** button from the **App Explorer** tab, we should see our application with two tabs. We can navigate from one window to the other using the tabs at the bottom of the screen (these tabs will be located at the top of the screen on Android), as shown in the following screenshot:

Appcelerator Cloud services

The official ACS documentation describes **Mobile Backend as a Service** (**MBaaS**). Simply put, it is a cloud provider that offers around 20 pre-built services that we find in most mobile applications these days. Such a service frees us from developing (and maintaining) our own backend service, thus saving time and effort.

At the time this book was written, ACS provided the following services:

Service	Description
Users	This service enables our users to create unique identification passwords and associate them with their first and last name for display to other users of our application (also used for access and authentication for our application).
Photos	This service enables our users to take photos from within our application, and then store and share them using the Cloud.
Custom Objects and Search	This service creates searchable data fields within our application. Basically, any custom JavaScript object can be stored and then searched depending on certain criteria.
Push Notifications	This service sends messages to any device without checking whether our application is currently running or not.
Email Templates	This service uses prebuilt templates to send e-mails from within our mobile application (customers can mail directly for support; employees can use e-mail templates to send expense reports, and many others).
Key Values	This service stores and retrieves string or even binary data up to 2 MB. This service offers a public store that can be viewed by every user for a given application. But each registered user also has a private key-value storage (note that a key-value pair can only be updated or deleted by its owner).
Places	This service uses location-based services to better address our user's needs (users can use Places to find the nearest restaurants or show advertisements related to their region).
Status	This service is used to make our application more social by enabling users to post status updates from wherever they are.

Service	Description
Posts	This service gives our users the ability to post updates, reviews, or other content, and allow others to respond to their posts.
Clients	This service understands what sorts of devices (brand, model, version) people are attaching to our application, so we can better adapt our content for the devices they are using.
Social Integration	This service uses the power of our user's social networks within our application by letting them log in and authenticate on Facebook and Twitter.
Checkins	This service allows our users to check in to a place or an event.
Chat	This service embeds instant messaging within our application to allow multiple users to interact directly.
Photo Collections	This service files any number of photos into a single album and then stores, shares, and publishes them within our application.
Ratings, Reviews, and Likes	This service adds ratings, reviews, and likes to our mobile application without having to build an entire rating machine and server-side infrastructure (customers can provide feedback about a product or service and even share their opinion with other users).
Access Control Lists	This service configures fine-grained control over who can see and/or write any type of object from ACS (for example, we can allow multiple users to share and update a collection of photos).
Events	This service creates and manages events (one time or recurring).
Files	This service uploads, manages, and distributes files up to 25 MB from our applications.
Friends	This service creates relationships between our users. An approval process can enforce these relationships, or they can be more open like followers.
Messages	This service sends and receives e-mail style messages between the users in our application.

This is a very detailed set of features that covers most of what is required in a modern, connected mobile application today.

ACS provides an additional service called Appcelerator node ACS that enables developers to publish Node.js applications to the cloud. It provides a fully-featured, command-line interface as well as its very own MVC framework to easily develop applications. It is also fully integrated inside Titanium Studio. This service can be used to build custom services for our applications. It can even be used to host our existing Node.js applications or services in the Appcelerator Cloud.

Creating a user

The first thing we need to do if we want to interact with ACS, is create a user. The reason behind this is that everything that is stored into ACS must be linked to an account. This approach makes perfect sense for an application where every user has his or her photos, messages, files, friends, and likes.

But for our application, we want to use Cloud as the central repository of information shared among every user that uses it. Since ACS cannot store any information without an account, we will simply create one user account for the application itself. Simply prove, as far as ACS is concerned, that every interaction is made by the same user.

We can access the Cloud using the available REST API that is accessible from any networked device for interacting with ACS objects.

REST (Representational state transfer) is a simple, stateless architecture that generally runs over HTTP (virtually in all instances, the HTTP protocol is used to create, read, update, or delete information). In many ways, the World Wide Web itself, based on HTTP, can be viewed as a REST-based architecture.

This is how we can create our user account using the `curl` command-line tool:

```
$ curl -b cookies.txt -c cookies.txt -X POST \
--data-urlencode "username=com.packtpub.hotshot.marcopolo" \
--data-urlencode "first_name=Marco" \
--data-urlencode "last_name=Polo App" \
--data-urlencode "password=12345" \
--data-urlencode "password_confirmation=12345" "https://api.cloud.
appcelerator.com/v1/users/create.json?key=<YOUR APP API KEY>"
```

Let's go over this last command in order to have a better understanding of what is actually sent to the server:

- The first line basically defines cookie parameters as well as the fact that we want to post the data to the server.

- Then, we define each attribute required for the creation of a user account:

    ○ The username, which will act as a login parameter; we will be using it every time we want our application to authenticate to the service (in our case, we used the application's ID).

    ○ The first and the last name (these can be anything, the values we may choose).

    ○ The desired password for this account as well as the confirmation, just to be sure.

- The last line is basically the URL to call, in order to create our account with our API key as the parameter.

> Remember that you will have to create two different user accounts; one with your development API key and a second one with your production API key.

Our development sandbox

We now have a working application and an ACS user account to which our data can be stored. We can now go ahead and interact with the cloud. The first thing we will do is load the `ti.could` module. At the very top of the `app.js` file, we will load the module using the `require` function and store its reference into the `Cloud` variable. Since our application has a single context, it will be accessible from everywhere in the code.

```
var Cloud = require('ti.cloud');
```

Another important thing to do while developing our application is to keep a clear separation between development data and real-life production data. While this may not apply at this stage of the development process, it shall become critical once the application is available to customers. For this, the module provides a `debug` property that we can set to `true` when we want to use our development API key. Once the development is complete, we just set this property to `false` or delete this line altogether:

```
Cloud.debug = true;
```

Connecting to the cloud

In our `app.js` file, we will create a new function named `loginAppUser` that will basically authenticate our application to ACS using the account we created earlier:

```
function loginAppUser() {
```

To authenticate a user account, we will call the `Users.login` function of the `Cloud` module. This function expects two parameters. The first parameter is an object containing the credentials (application ID and password). The second parameter is a callback function that will be invoked when the service sends its response:

```
Cloud.Users.login({
    login: Ti.App.id,
    password: '12345'
}, function (e) {
```

If successful, we will extract the returned `user` object from the response and log its attributes to the console.

```
if (e.success) {
    var user = e.users[0];

    Ti.API.info('Success:\n' +
        'id: ' + user.id + '\n' +
        'sessionId: ' + Cloud.sessionId + '\n' +
        'first name: ' + user.first_name + '\n' +
        'last name: ' + user.last_name);
```

If we come across an error, we will display an alert containing the error code and message. If there is no error message (for example, network error), we will display the error object in string form:

```
} else {
    alert('Error:\n' +
        ((e.error && e.message)
        || JSON.stringify(e)));
    }
    });
}
```

This function is quite basic and it can have additional features, but it is enough for a simple instance such as ours.

Don't forget to call it

Since we want our application to authenticate to ACS every time it starts, we will invoke a newly created function in our self-calling function. Now, the authentication will be the very first action taken by our application:

```
(function() {
  loginAppUser();

  ...
  // Rest of our code
})();
```

Just to make sure

We will test the application one more time to verify the authentication process. Unless we are shown an alert on the screen, there is no visual indication to verify whether it actually worked. Our only indication is in the console where we should see the following information:

```
[INFO] :    Success:
[INFO] :    id: 51fab18b81df773a5d06cdd5
[INFO] :    sessionId: 7OgkYGcMJXITacsnFEHtiR1PkUk
[INFO] :    first name: Marco
[INFO] :    last name: Polo App
```

The Polo window

Implementing a reverse approach to the Marco Polo concept, we will first give our players the ability to provide information about their position to other players using the cloud. This approach makes sense because there is a little chance of getting the other player's location if no one shares it first.

The user interface

We will edit our `PoloWindow.js` file and add the following code right after where we created our window. We will begin by adding a new label at the top:

```
self.add(Ti.UI.createLabel({
  top: 17,
  width: '80%',
  height: Ti.UI.SIZE,
  color: '#000',
  text: 'Enter your name'
}));
```

We will then create a text field that will require our players to enter their name.
It will span 80 percent of the screen's width, having a solid border with rounded
corners. For its value, we will rely on a property named PLAYER_NAME so that
it persists even when the player quits the application. If this is the first time the
application is started (meaning there is no property saved yet), we will use an empty
string as a value. Finally, we will save its reference in the txtPlayerName variable
and then add it to our window:

```
var txtPlayerName = Ti.UI.createTextField({
  top: 40,
  width: '80%',
  borderStyle: Ti.UI.INPUT_BORDERSTYLE_ROUNDED,
  value: Ti.App.Properties.getString('PLAYER_NAME', '')
});


self.add(txtPlayerName);
```

After this, we will create our **Polo** button. We will use a standard view and give it a
green background. We will also give it a thick white border and set its borderRadius
attribute to 100% so that it appears as a filled circle. We will store its reference in the
btnCheckin variable since we want to click on it later:

```
var btnCheckin = Ti.UI.createView({
  width: 200,
  height: 200,
  backgroundColor: '#8ca93e',
  borderColor: '#fff',
  borderWidth: 6,
  borderRadius: '100%'
});
```

Right in the middle of our circular button, we will add a white label with a big
bold font:

```
btnCheckin.add(Ti.UI.createLabel({
  width: Ti.UI.FILL,
  height: Ti.UI.SIZE,
  color: '#fff',
  text: 'Polo',
  textAlign: Ti.UI.TEXT_ALIGNMENT_CENTER,
  font: {
    fontSize: '45sp',
    fontWeight: 'bold'
  }
}));
```

As with every other control, we need to add it to our window:

```
self.add(btnCheckin);
```

The very last user interface control for our window will be a label that we will position at the very bottom of the screen to display the Geolocation status. It will occupy the entire width of the screen and the text will be centered. It will also have a purple background, but we will not make it completely opaque for a nicer look. As always, we will need to add it to our window:

```
var lblStatus = Ti.UI.createLabel({
  bottom: 0,
  width: Ti.UI.FILL,
  backgroundColor: '#6f0564',
  color: '#fff',
  opacity: 0.7,
  textAlign: Ti.UI.TEXT_ALIGNMENT_CENTER
});

self.add(lblStatus);
```

One more thing

As with every application that uses text fields, we must make sure we hide the keyboard once our players are done editing the field. The easiest way to achieve this is to call the `blur` function when the player taps anywhere on the screen:

```
self.addEventListener('click', function() {
  txtPlayerName.blur();
});
```

Determining the player's location

With the completion of the Polo window, we now need to determine the player's current position before we can upload it to the Cloud. In order to hide the Geolocation code from the ACS interaction code, we will isolate Geolocation in its own file. Also, both of our application's windows will require access to the Geolocation service. This approach will avoid duplicate our code in each window.

A better Geolocation service

For this, we will create a new file named `GeolocationService.js` and place it into a new `Resource/service` directory.

In this new file, we will create a new function named `findMe`, which will determine the player's location. It will expect a `callback` function as parameter, which will be invoked once the device's location is determined. The reason for this, is that getting the device's location is an asynchronous process, meaning it may not be completed by the time we reach the end of the function. By using a callback, we will be assured that the process is complete:

```
function findMe(callback) {
   var status, lat, lon;
```

If the Geolocation service is activated on the device, we will give it a purpose (so our players knows why we want to get the location). Also, we set the accuracy to the highest possible setting:

```
if (Ti.Geolocation) {
   Ti.Geolocation.purpose = 'To find current location.';
   Ti.Geolocation.accuracy = Ti.Geolocation.ACCURACY_BEST;
```

Once the current position has been retrieved, we will check whether the service has returned any error whatsoever. If yes, we will assign the `status` message accordingly. On the other hand, if the location was successfully determined, we will update the `lat` (latitude) and `lon` (longitude) variables with the device's coordinates, as well as the `status` message:

```
Ti.Geolocation.getCurrentPosition(function (e) {
   if (!e.success || e.error) {
      status = 'GPS lost';
   } else{
      status = 'Location acquired';

      lat = e.coords.latitude;
      lon = e.coords.longitude;
   }
```

We will then invoke the `callback` function with a dictionary object as the parameter containing the `status` message, as well as the geographic coordinates:

```
callback({
   status: status,
   longitude: lon,
   latitude: lat
});
});
```

There might be cases where the device's Geolocation service won't be active. Therefore, we will rely on another way to determine the location provided by ACS. It uses the IP address of the device and then relies on a service called MaxMind GeoIP to return the most accurate IP-based Geolocation data possible. Note that the results are not based on GPS signals or even Wi-Fi triangulation. It actually matches your IP address with a known physical location (for example, a restaurant). While this will not be as accurate as a true GPS, it can prove to be a good backup solution in big connected cities.

We will use the `Clients.geolocate` function to get the device's location. If the operation is successful, we will update the `lat`, `lon`, and `status` variables much similar to what we did previously:

```
} else {
  Cloud.Clients.geolocate(function (e) {
    if (e.success) {
      status = 'Location acquired';

      lat = e.location.latitude;
      lon = e.location.longitude;
    }
```

If any error is returned by the ACS Geolocation service, we will update the `status` message accordingly to inform the player:

```
else {
  status = 'GPS lost';
}
```

Just as we did for the standard Geolocation service, we will invoke the `callback` function with the `status` attribute and coordinates as parameters:

```
callback({
  status: status,
  longitude: lon,
  latitude: lat
});
    });
  }
}
```

Finally, we need to export the function in order to make it accessible outside the module:

```
exports.findMe = findMe;
```

Pushing our location to the Cloud

Now that we know how to retrieve our device's location, we will go ahead and push this same location to the cloud. To do this, we will use the Places service provided by ACS. Each player can only be at one place at any given time; therefore, when our players will select the **Polo** tab on the screen, we will create a place object with its location data and push it to the cloud. Since we want each place (player on the map) to be unique, we will also assign the player's name to our object so that we can distinguish them later on in the game.

 At first glance, it would seem that the Checkins service might just be the right fit for our needs, so why are we using the Places service instead? The reason is quite simple; Checkin objects in ACS must be an actual place and not precise coordinates on the map. This is good when you want your users to check-in at the library or a coffee shop. But for a precise location on the map, only Places offers us the flexibility we need.

At the very top of our `PoloWindow.js` file, we will load our newly created Geolocation service and keep its reference in a variable named `GeolocationService` for later use:

```
var GeolocationService = require('service/GeolocationService');
```

We will then create a new function named `pushToCloud`, which will be invoked when the device's location has to be determined. It will only have one parameter named `geo` and will contain the device's location as well as the status message.

First, we will update the text of the status label with the message returned by the service:

```
lblStatus.text = geo.status;
```

We mentioned earlier that every object from ACS has an ID, and Places are no exception to the rule. Our approach consists of having one player in one place at any given time. Therefore, our players become places themselves, each having their very own ID. This is why we will store that ID onto our device for every later operation. We will retrieve the property and store its value in the `placeId` variable. If the property is not found for some reason, we will return an empty string:

```
var placeId = Ti.App.Properties.getString('PLACE_ID', '');
```

This is our very first time

If there is no place ID saved on the device, it usually means that the player has never pushed his or her location to the Cloud yet. In this instance, we will then create a new place object using the `Places.create` function of the `Cloud` module. This function takes two parameters. The first one is a dictionary object, which contains the name of the place (the player's name) and the coordinates. The second parameter is a callback function that will be invoked when the service responds:

```
if (!placeId) {
  Cloud.Places.create({
    name: txtPlayerName.value,
    latitude: geo.latitude,
    longitude: geo.longitude
}, function(e) {
```

If the creation was successful, we will extract the first place object from the response (our newly created place object). We will then save its ID under the `PLACE_ID` property, which can be used when the player wants to update his/her location again. Of course, we will also update the status label to inform the player that the creation was successful. If the creation process fails, we will log an error to the console:

```
if (e.success) {
  var place = e.places[0];

  Ti.App.Properties.setString('PLACE_ID', place.id);

  lblStatus.text = 'Position saved to Cloud!';
} else {
  Ti.API.error(msg);
}
});
```

We already played this game before

If a place ID is already stored on the device, it means that the player has already provided his first update to the Cloud. This means that we need to update the already existing object instead of creating a new one.

What we will do here is quite similar to the creation process. We will call the `Places.update` function from the `Cloud` module. We will pass a dictionary object containing the ID of the place object we want to update as well as all the information we want to update (all fields are optional, except the ID). If the update is successful, we will update the status label to inform the player. If it fails, we will log an error to the console:

```
    } else {
      Cloud.Places.update({
        place_id: placeId,
        name: txtPlayerName.text,
        latitude: geo.latitude,
        longitude: geo.longitude
      }, function(e) {
        if (e.success) {
          lblStatus.text = 'Cloud position updated!';
        } else {
          Ti.API.error(msg);
        }
      });
    }
  });
```

We will then add an event handler to our big round button when the player taps on the button. The very first thing we will do in this handler will be to make sure that the player has entered a valid name in the text field. If that is not the case, we will simply show an alert asking him or her to do so, and then quit the function:

```
btnCheckin.addEventListener('click', function(e) {
  if (txtPlayerName.value.length < 3) {
    alert('Please enter a valid name');
    return;
  }
```

Since we want the username to be persistent between sessions, we will save its value under the PLAYER_NAME key as follows:

```
    Ti.App.Properties.setString('PLAYER_NAME', txtPlayerName.value);
```

Even if the player wants to change his or her name, it will not be a problem since every place object has a technical ID used to access its content. Therefore, our players can change the screen name at will.

We will change the text of the status label to inform our players that the `GeolocationService` variable will be called. Finally, we will invoke the `findMe` function to determine the device's location, and pass `pushToCloud` as a callback function to be invoked once `GeolocationService` locates the device:

```
lblStatus.text = Getting location, please wait...';
GeolocationService.findMe(pushToCloud);
});
```

Testing our Polo feature

We have looked at all the features of our first window and now we can launch our application and see it in action. When we launch the application and select the **Polo** tab, a window opens where we can enter our desired screen name and then upload our current location to the Cloud by tapping the big round button. We should see the status text changing at the bottom of the screen with every step of the way.

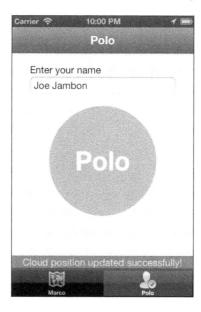

The Marco window

Now that our players can push their location to the Cloud, we can retrieve those same locations and display them on a map view using annotations (pins). There is not much code needed for the user interface, since the only component is a map view that will occupy the entire screen's real estate.

Creating a dedicated map module

Just as we did with the GeolocationService variable, we will isolate all the code related to mapview into its own file for more clarity. We will create a new file named Map.js into our Resources directory and create two functions in it.

The first function will be named createMap, which will return a new mapview variable. It will be of standard type, meaning that it will display a street map with street names, for example. It will use a smooth animation when the map region changes and will adjust its aspect ratio to fit the region using the regionFit property. Also, we want to show where the player is located on the map by setting the userLocation property to true:

```
function createMap() {
  var mapview = Titanium.Map.createView({
    mapType: Titanium.Map.STANDARD_TYPE,
    animate: true,
    regionFit: true,
    userLocation: true
  });

  return mapview;
}
```

The second function will return a map annotation (also called `pin`). This new function will expect one dictionary object as the parameter. It will contain the geographic coordinates on the map as well as on the `title` attribute to display when the player selects a pin on the screen. Every annotation will be the color red, to maintain the theme with the navigation bar, and they will appear on the map with nice animation:

```
function createAnnotation(params) {
  var pin = Ti.Map.createAnnotation({
    latitude: params.lat,
    longitude: params.lon,
    title: params.title,
    pincolor: Titanium.Map.ANNOTATION_RED,
    animate: true
  });

  return pin;
}
```

Finally, we will export both of the functions so they can be accessed from any code that loads our module:

```
exports.createAnnotation = createAnnotation;
exports.createMap = createMap;
```

We have CommonJS modules, let's use them

This new window needs to use Geolocation as well as a map view. Therefore, we will load both the modules at the very top of our `MarcoWindw.js` file:

```
var GeolocationService = require('service/GeolocationService');
var Map = require('Map');
```

We will then create our `mapview` variable inside the `MarcoWindow` function, right after the window is created.

```
var mapView = Map.createMap();
```

We will then add it our window's only control, so it is displayed on the screen:

```
self.add(mapView);
```

Getting player locations from the cloud

Now that we have set up everything for our map, we will use it to display the other player's locations. We will create a new function named `updateMap`, which will be invoked once the Geolocation service determines the device's location. It will also have one parameter named `geo`, which will contain the geographic coordinates as well as the status message:

```
function updateMap(geo) {
```

We will then invoke the `Places.search` function from the `Cloud` module. This function can search for any `Places` object at a specific location, which is within a certain radius. But for this particular case, we need to retrieve every single player's location. Therefore, now we will now just invoke the function without any parameters as follows:

```
Cloud.Places.search({}, function(e) {
```

If the search was successful, we will create an empty array aptly named `annotations`, which as its name states will contain every annotation variable we want to display on our map view:

```
if (e.success) {
  var annotations = [];
```

We will then loop through every `place` object returned by our query and create a new `annotation` object using the `place` object's coordinates and name. We will also add each newly created `annotation` variable to the `annotations` array:

```
for (var i = 0; i < e.places.length; i++) {
  var place = e.places[i];
  annotations.push(Map.createAnnotation({
    lat: place.latitude,
    lon: place.longitude,
    title: place.name
  }));
}
```

We will then assign our `annotations` array to our `mapview` variable in order to display them:

```
mapView.setAnnotations(annotations);
} else {
```

If there occurred any error during the ACS call, we will display an alert on the screen to inform the player. If the service returns an error code or message, we will display it. If the error is something other than ACS, we will display the error object as a string:

```
alert('Error:\n' +
   ((e.error && e.message)
   || JSON.stringify(e)));
}
});
```

Finally, we will set the map region we want to display. In our case, we will center the map to the device's current location; also, we will set the latitude and longitude delta to 0.75 degrees, which will zoom out the map enough to show the surrounding cities. Finally, we would want the `mapView` component to have a smooth animation while zooming to the specified region:

```
mapView.setRegion({
   latitude: geo.latitude,
   longitude: geo.longitude,
   latitudeDelta:0.75,
   longitudeDelta:0.75,
   animate: true
});
}
```

We will create an event handler to be triggered after the window opens. We will then retrieve the device's current location by calling the `findMe` function, without forgetting to pass the `updateMap` function as a callback to be invoked once the location is determined:

```
self.addEventListener('open', function() {
   GeolocationService.findMe(updateMap);
});
```

Let's play!

With our application feature now complete, we can take it for one last spin. While launching the application, we should see the map zooming to the region where the device is located. We should then see that the pins are being added onto the map. If we were to tap on one of the pins, we should see a description appearing over it displaying the selected player's name, as shown in the following screenshot:

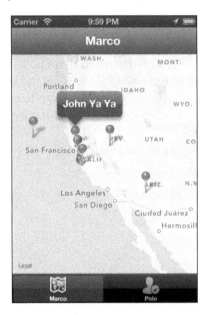

What about Android?

While trying to run our application on an Android emulator or device, it might work without any problem. But it is important to note that as of March 3, 2013, Google deprecated the first version (v2) of their Google Maps API. In fact, Google will not issue any more v1 API keys, thus pushing developers to its new version (v2).

To address this, Titanium introduced a new native module called `Modules.Map`, which replaces the old `Titanium.Map` module for the Android platform. This module offers better performance, provides a lot of new features and the ability to add more than one map in our application (such a feat was not possible on Android with the previous version).

Before making any changes to our project, the following are the steps we must perform:

1. Install the Google Play services SDK using the Android SDK manager.
2. Obtain a Google API key from Google's developer website.
3. Add this new API key to the `<android>` section of our `tiapp.xml` file.

These steps are out of the scope of this chapter, but they are explained in great detail in the online Titanium documentation (refer to the *Appendix*, *References*).

Once everything is in place, we can go ahead and integrate the new module into our project. We will add a reference to the `ti.map` module in our `tiapp.xml` file:

```
<modules>
  <module platform="commonjs">ti.cloud</module>
  <module platform="android">ti.map</module>
</modules>
```

It's good that maps are their own thing

Earlier in the project, we isolated all of the code related to the map view in its own file named `Map.js`. This will make it very easy to have a new specific implementation for the Android platform.

First, we will move our existing `Map.js` file from our `Resources` directory to the `iphone` subdirectory. Second, we will create a brand new file named `Map.js` in the `android` subdirectory as well; this file will contain the same functions (with the same names), so it will be completely transparent to the code calling this module. Just as with images, Titanium will use the appropriate file while building the target platform.

In our newly created file, we will first load the `ti.map` native module and save its reference into the `MapModule` variable as follows, since we will be using it later:

```
var MapModule = require('ti.map');
```

We will then implement the `createMap` function that will return a map view created using the new native module:

```
function createMap() {
  var mapView = MapModule.createView({
    mapType: MapModule.NORMAL_TYPE,
    animate:true,
    regionFit:true,
    userLocation:true
  });

  return mapView;
}
```

We will do the same for the `createAnnotation` function, which will return a red annotation at the desired geographic coordinates:

```
function createAnnotation(params) {
  var pin = MapModule.createAnnotation({
    latitude: params.lat,
    longitude: params.lon,
    title: params.name,
    pincolor: MapModule.ANNOTATION_RED,
  });

  return pin;
}
```

As we did with the iOS implementation, we will export our two functions so they are accessible from the outside world:

```
exports.createAnnotation = createAnnotation;
exports.createMap = createMap;
```

Testing this new map module

Before we test our new Android implementation, there are certain limitations we must keep in mind. For example, this new module can only be tested on an actual physical device. The new module is not compatible with the Android emulator.

Also, Google Play services must be installed on the device. If that is not the case, we will be confronted with one of the following outcomes:

- The map view is completely black
- The system displays an alert message stating that Google Play services is missing
- The application crashes altogether

Maps cannot work without Google Play Services; we can, however, verify whether the services are installed and then notify our players with a more appropriate message using the isGooglePlayServicesAvailable function.

This function is provided by the module.map and can return one of the following outputs:

Constant	Description
SUCCESS	This constant displays that Google Play services is installed.
SERVICE_MISSING	This constant displays that Google Play services is missing. Users need to install it from the Google Play store.
SERVICE_VERSION_UPDATE_REQUIRED	This constant displays that Google Play services is out of date, indicating that users need to update their devices.
SERVICE_DISABLED	This constant displays that Google Play services is disabled. Users must enable Google Play services on their devices if they want to use the new Google Maps.
SERVICE_INVALID	This constant displays that Google Play services cannot be authenticated, indicating that users need to reinstall Google Play services on their devices.

Summary

This final chapter walked us through the development of an application that used Appcelerator Cloud Services to share the device's geographic location with other people. We learned to use tab groups to give our players the ability to navigate from one window to another with ease. We learned that ACS also provides Geolocation features, which we can use if we haven't activated the GPS service on our device. We also learned how to use a map view in more detail, and also how to add pins on to the map. Finally, we used the latest version of the Google Maps component for better performance, and also to be future proof (since Google has now deprecated its previous version).

References

The source code for this book

All the code covered in this book is freely available on the author's public GitHub repository. It is open source and is under the Apache License, Version 2.0. This means that anyone can clone the repository, modify the code, and re-use it to fit their needs. It contains the code for 10 complete applications as well as the necessary assets.

It contains the Stopwatch library for *Chapter 1*, *Stopwatch (with Lap Counter)*, all the native extension modules used for each application, media assets (images, icons, video, SpriteSheets, and tilesets), and many more.

The repository can be found at the following URL:

```
https://github.com/TheBrousse/TitaniumMobileHotshot
```

The Page Flip Module

The Page Flip Module is used in *Chapter 4*, *Interactive E-Book for iPad*. This module was developed by Appcelerator and can be downloaded for free from the Appcelerator Marketplace at `http://bit.ly/1bykJjd`.

This module is open source and is under the Apache license, Version 2.0. Its complete source code can be found at the following URL:

```
http://bit.ly/1eLGqfo
```

The cURL utility

cURL (**Client for URLs**) means that it is a tool that allows users to interact with URLs from the command line. There are versions available for every platform supported by Appcelerator Titanium, and it is often preinstalled on most operating systems.

If that is not the case (for example, in Windows), it can be downloaded from the official cURL website, `http://curl.haxx.se/`.

The Stock quote API

In *Chapter 5, You've Got to Know When to Hold 'em*, all the stock prices are retrieved using a public web service provided by a company named Market On Demand. Their API provides useful functions related to the financial market, such as `Company Lookup`, `Stock Quotes`, and `Timeseries` (used to determine the historical value of a company's stock).

The complete API documentation is available on their developer website at `http://dev.markitondemand.com`.

The tiled map editor

Tiled is a general-purpose tile map editor. It is meant to be used for editing maps of any tile-based game, be it an RPG similar to what we saw in *Chapter 6, JRPG – Second to Last Fantasy*, or any other type of game. It is mostly written in C++ using the Qt libraries, which means that it is compatible on most platforms. It can be downloaded for free from their official website at `http://www.mapeditor.org`.

The source code is completely open source and is under two different licenses (GPL and Simplified BSD depending on what code you are modifying) and can be found in GitHub at `https://github.com/bjorn/tiled`.

Sprite sheets and tilesets

All graphical assets used for both *Chapter 6, JRPG – Second to Last Fantasy*, and *Chapter 7, JRPG – Second to Last Fantasy Online*, have been drawn by a gentleman named *Seth Jester*. His site is full of characters, objects, and tilesets, and he makes them available for free at `http://bit.ly/crBh2T`.

All that he asks in return is that his work should be credited, which is a small price to pay considering the level of quality of his work.

The QuickTiGame2d game engine

This is the game engine used in *Chapter 6, JRPG – Second to Last Fantasy*, and *Chapter 7, JRPG – Second to Last Fantasy Online*. It is a native extension module that can be used to access low-level graphics function. It allows developers to develop graphic-demanding applications such as games. The module is open source and is under the new BSD license. However, the repository was closed on October 2012 when Appcelerator invested in a start-up named Lanica (`http://lanica.co`) that now offers the evolution of that module as a commercial offering.

Although the source code is not available anymore, the engine itself is still available for download at the following URLs:

File Description	Download Link
QuickTiGame2d Module 1.2 for iOS	`http://bit.ly/Z8YsEw`
QuickTiGame2d Module 1.2 for Android	`http://bit.ly/11rDRs7`

They also provide Wiki-based documentation at `http://bit.ly/19k4pBW`.

Node.js

Node.js is a software platform that is used to develop (usually) server-side applications. It is built on Google Chrome's JavaScript runtime; therefore, developers can use JavaScript on the server. It is also important to mention that it uses a single-threaded event loop and nonblocking I/O, which provides very high performance. It can be downloaded for free from the official website at `http://nodejs.org`.

NPM

Node Package Manager (**NPM**), just as its name implies, is the predominant package manager for the Node.js platform. There were more than 38,000 packages available in the npm registry (`https://npmjs.org`) at the time of writing this book. It comes bundled with the Node.js installation, so no operation is needed in order to use it.

Socket.IO

`Socket.IO` is a JavaScript library for real-time web applications. It has both a server component as well as a client component. It then uses the web socket protocol for communication. This means that it relies on events instead of always soliciting the server to check whether it has some new information available.

It can be installed using the npm tool, making it easy to be set up.

The TiWS module

TiWS is a native extension module used for creating native web sockets and was developed by a gentleman named *Jordi Domenec*. It is possible to use it with `Socket.IO` and other popular libraries. This is the reason why it was chosen in *Chapter 7, JRPG – Second to Last Fantasy Online*, for the part that dealt with communication. It can be downloaded from the Appcelerator Marketplace. Some of the downloadable files are given in the following table:

File Description	Download Link
TiWS module 0.3 for iOS	`http://bit.ly/1529TwA`
TiWS module 0.1 for Android	`http://bit.ly/142HLKG`

This source code is open source and is under the Apache License, Version 2.0. It is available on GitHub at `http://bit.ly/L8eFzj`.

The Facebook Graph API

The Graph API is literally the core of the Facebook platform. It is the primary way to retrieve or post data to Facebook. It represents a consistent view of objects contained in the graph (people, photos, events, and pages) and the connections between those objects (friendships, shared content, photo tags, and many more).

All the functions provided by the API is thoroughly documented on the Facebook developers' website:

`http://developers.facebook.com/docs/reference/api/`

The social_plus library

The application covered in *Chapter 8, Social Networks*, interacts with Twitter. The `social_plus` library provides an easy OAuth authentication as well as a function to publish status messages and photos on Twitter. The library is open source and is available on GitHub. `http://bit.ly/ThUK67`.

A gentleman named *Aaron Saunders* developed it. *Aaron* was one of the first advocates of the Titanium framework and has been a valued member of the community since the very beginning.

The Flickr API

The Flickr Photo service provides an API that offers a large array of functions to add, update, or delete information from their system. This API is available for noncommercial use by outside developers. Commercial use is possible by prior arrangement with the website.

All the detailed information regarding each function is contained on their developer website at `http://www.flickr.com/services/api/`.

Appcelerator Cloud Services

Appcelerator Cloud Services (**ACS**) provides around 20 prebuilt services that are used in most web or mobile applications today (Status, Check-ins, Chat, Photos, and a lot more). They provide an SDK that is already embedded into the Titanium framework, and they also provide dedicated SDKs for iOS (to use with Xcode) as well as for Android. But ACS is not only targeted at mobile applications; any application can pretty much interact with ACS using a REST API that they provide.

Each of these SDKs is detailed in the official Titanium documentation at `http://docs.appcelerator.com/cloud/latest/`.

The MaxMind GeoIP service

The application covered in *Chapter 10, Worldwide Marco Polo*, determines the device's location using the embedded GPS. But for cases where the Geolocation service is not available, the application falls back on a function that relies on a service called MaxMind GeoIP. This service enables developers to identify the location, organization, connection speed, and user type based on the device's IP address.

All the documentation regarding this service can be found on their website at `http://bit.ly/UJGkdQ`.

Google Maps v2

On March 3, 2013, Google introduced their new Google Map API v2, thus deprecating the previous version (Google Maps API v1). Since Google will not issue any more API keys for the earlier version, it is recommended that every new application should use the new version. Some of the main reasons are performance improvement and the ability to have more than one map view into an application.

While *Chapter 10, Worldwide Marco Polo*, covers the coding aspect, it is essential to have an API key in order to use the new map module. The API key can easily be obtained on Google's Android developer website at `http://bit.ly/10kkTsU`.

Index

Thank you for buying
Creating Mobile Apps with Appcelerator Titanium

About Packt Publishing

Packt, pronounced 'packed', published its first book "*Mastering phpMyAdmin for Effective MySQL Management*" in April 2004 and subsequently continued to specialize in publishing highly focused books on specific technologies and solutions.

Our books and publications share the experiences of your fellow IT professionals in adapting and customizing today's systems, applications, and frameworks. Our solution based books give you the knowledge and power to customize the software and technologies you're using to get the job done. Packt books are more specific and less general than the IT books you have seen in the past. Our unique business model allows us to bring you more focused information, giving you more of what you need to know, and less of what you don't.

Packt is a modern, yet unique publishing company, which focuses on producing quality, cutting-edge books for communities of developers, administrators, and newbies alike. For more information, please visit our website: www.packtpub.com.

About Packt Open Source

In 2010, Packt launched two new brands, Packt Open Source and Packt Enterprise, in order to continue its focus on specialization. This book is part of the Packt Open Source brand, home to books published on software built around Open Source licences, and offering information to anybody from advanced developers to budding web designers. The Open Source brand also runs Packt's Open Source Royalty Scheme, by which Packt gives a royalty to each Open Source project about whose software a book is sold.

Writing for Packt

We welcome all inquiries from people who are interested in authoring. Book proposals should be sent to author@packtpub.com. If your book idea is still at an early stage and you would like to discuss it first before writing a formal book proposal, contact us; one of our commissioning editors will get in touch with you.

We're not just looking for published authors; if you have strong technical skills but no writing experience, our experienced editors can help you develop a writing career, or simply get some additional reward for your expertise.

Appcelerator Titanium Business Application Development Cookbook

ISBN: 978-1-849695-34-3   Paperback: 328 pages

Over 40 hands-on recipes to quickly and efficiently create business grade Titanium Enterprise apps

1. Provide mobile solutions to meet the challenges of today's Enterprise mobility needs

2. Study the best practices in security, document management, and Titanium Enterprise Development

3. Create cross-platform Enterprise class Titanium apps quickly and efficiently with step-by-step instructions and images to help guide you

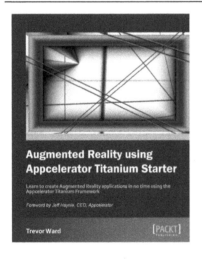

Augmented Reality using Appcelerator Titanium Starter

ISBN: 978-1-849693-90-5   Paperback: 52 pages

Learn to create Augmented Reality applications in no time using the Appcelerator Titanium Framework

1. Learn something new in an Instant! A short, fast, focused guide delivering immediate results.

2. Create an open source Augmented Reality Titanium application

3. Build an effective display of multiple points of interest

Please check **www.PacktPub.com** for information on our titles

Appcelerator Titanium Application Development by Example Beginner's Guide

ISBN: 978-1-849695-00-8   Paperback: 334 pages

Over 30 intresting recipes to help you create cross-platform apps with Titanium, and explore the new features in Titanium 3

1. Covers iOS, Android, and Windows8

2. Includes Alloy, the latest in Titanium design

3. Includes examples of Cloud Services, augmented reality, and tablet design

Appcelerator Titanium: Patterns and Best Practices

ISBN: 978-1-849693-48-6   Paperback: 110 pages

take your Titanium development experience to the next level, and build your Titanium knowledge on CommonJS structuring, MVC model implementation, memory management, and much more

1. Full step-by-step approach to help structure your apps in an MVC style that will make them more maintainable, easier to code and more stable

2. Learn best practices and optimizations both related directly to JavaScript and Titanium itself

3. Learn solutions to create cross-compatible layouts that work across both Android and the iPhone

Please check **www.PacktPub.com** for information on our titles

www.ingramcontent.com/pod-product-compliance
Lightning Source LLC
Chambersburg PA
CBHW080353060326
40689CB00019B/3994